LIVING BIOLOGY IN SCHOOLS

edited by

Michael Reiss

Institute of Biology

Institute of Biology
20-22 Queensberry Place, London, SW7 2DZ, United Kingdom.

© 1996 Institute of Biology

Edited by Revd Dr M J Reiss *CBiol FIBiol*

Education Working Party:
Catherine Bennett, Ian Carter, Michael Cassidy, Nigel Collins, Jane Goring, Rachel Gravett, Michael Reiss, Barbara Tomlins.

Editorial Secretariat:
Anne Jordan (Head of Education Department)
Jonathan Cowie (Head of Books & Sectors of Biology)

Library Cataloguing Data

Dewey (20th ed) 375.574

Living biology in schools

1. Teaching
2. Biology
I. Title

ISBN 0-900490-32-2

Printed in the United Kingdom by Birbeck & Sons, Birmingham.

Contents

Preface		1
1.	**Introduction** R.Gravett	3
2.	**Microbiology** J.Schollar and J. Grainger	7
3.	**Plants** N.Collins and R. Price	29
4.	**Animals** M.Cassidy and J.Tranter	49
5.	**Pupils as a resource** P.Horton	67
6.	**Legalities** C.Bywater	77
7.	**Safety** M.Ingram	95
8.	**Moral and ethical issues** R.Lock and M.Reiss	109

PREFACE

Good biology has always introduced pupils and students to the variety of living organisms. However, for a number of reasons, recent years have shown a decrease in the use of living organisms in schools. Some of the thinking behind this decrease is to be applauded: teachers, students and pupils are now more aware of the dangers of causing suffering to animals or of damaging the environment. Other reasons reflect the varied pressures faced by teachers - budgetary constraints, shortage of time and stricter controls over what is and what is not permitted in classrooms.

In 1992 the Institute of Biology therefore set up a Working Party to address the issues surrounding the use of living organisms in schools. It was decided that we would work towards producing a short book to update and replace an early IOB publication "Living Biology in the Classroom". After considerable discussion, we arrived at the following aims and objectives:

1. To promote the use of living organisms in schools.
2. To provide clear guidance on the legal issues surrounding the use in schools of microorganisms, plants and animals.
3. To inform on safety aspects of the use of living organisms in schools.
4. To provide guidance on the use of pupils as subjects.
5. To explore moral and ethical concerns connected with the use of organisms in school teaching.
6. To provide resource and reading lists.

We have not tried to tackle fieldwork, but have confined ourselves to classroom teaching across the 5 to 19 range. Although we do not wish to be too parochial, we have sought to demonstrate that excellent teaching in this area can be used to fulfil many of the requirements of the Science National Curriculum (England and Wales).

We have enjoyed producing this book and feel we have learned much in the process. We hope that it will be widely read by primary teachers, secondary science teachers, science technicians, LEA officers and science educators. Our fundamental hope, of course, is that pupils may enjoy learning biology through the provision of first rate teaching based on contact with the whole range of living organisms.

Catherine Bennett, Ian Carter, Michael Cassidy, Nigel Collins, Jane Goring, Rachel Gravett, Michael Reiss, Barbara Tomlins

Chapter 1 INTRODUCTION

R. Gravett

Biology is the study of Life so it would seem vital that living things are regarded as an integral part of the study of the subject.

The lives of plants and other animals are intricately linked with our own. Apart from obvious ways that we 'use' them - plants as a source of food, building materials, paper, shade and decoration; animals as food, for work, for scientific purposes, in entertainment, in sport and for companionship - it is only recently that the human race is beginning to recognise quite how dependent it is on the grand network of living things. This recognition is exemplified by the many references to living organisms in the Science National Curriculum (England and Wales) (Department of Education 1995).

Attainment Target 2 of the Science National Curriculum, Life Processes and Living Things and associated programmes of study, makes clear the value of contact with, and studies involving, plants and animals. At Key Stage 1 it states that 'Pupils should be taught that animals, including humans, move, feed, grow, use their senses and reproduce' and 'that plants need light and water to grow'. It goes on to say 'Pupils should be given opportunities to consider how to treat living things and the environment with care and sensitivity' and pupils should be taught 'that there are different kinds of plants and animals in the local environment.'

At Key Stage 2, 'Pupils should be taught that there are life processes, including nutrition, movement, growth and reproduction, common to animals, including humans', 'that there are life processes, including growth, nutrition and reproduction, common to plants' and 'that microorganisms exist, and that many may be beneficial while others may be harmful'.

At Key Stage 3, 'Pupils should be given opportunities to relate scientific knowledge and understanding to the care of living things and of the environment'.

Finally, GCSE Biology includes among its aims: to develop a lifelong interest in, and enjoyment of, the study of living organisms and to promote a respect for all forms of life.

Practical work involving inanimate materials, such as tuning forks and chemicals, which do not have feelings, do not move, do not go to seed or grow out of control, present fewer complications than studies involving living things such as plants, animals, microorganisms or pupils themselves. Living material is less predictable and involves behind-the-scenes husbandry with careful planning and timing. Moral issues also influence studies of living things (usually animals) which, while presenting valuable learning opportunities, may be seen as yet another obstacle. A recent survey has been carried out to ascertain opinions of teachers on the use of living organisms in secondary school science (Reiss and Beaney, 1992). It was found that over the last few years several factors have constrained the use of living organisms in schools. These include the pressures of time, lack of resources, other effects of the introduction of the National Curriculum (England and Wales), a shortage of trained technicians and a shift in pupils' perceptions about the ethics and value of the use of animals in schools. Despite all these constraints, most respondents strongly favoured the appropriate use of living organisms as an integral part of biology education. It is hoped that this book will encourage and support teachers and technicians in the use of living organisms, when they wish to include them, by giving guidance on suitable species, ideas for practical work and clarifying the situation with regard to safety, legalities and the moral and ethical implications.

Chapter 1 Introduction

Much has been written to justify the inclusion of animals in biology teaching; somewhat less about plants because their use is not so controversial. Many of the arguments supporting the use of animals, however, apply to living things in general.

In a joint statement by the Association for Science Education, Institute of Biology and Universities Federation for Animal Welfare, the following educational benefits of keeping animals in schools are identified:

1. Opportunities for detailed observation and investigation of animal behaviour, structure and function, growth and life cycles.
2. Contributions to the social education of students through observation and discussion of reproduction, social interactions, and death, leading to an appreciation of the material and social needs of animals - including human beings.
3. Increased motivation to study living animals and to use and develop skills of literacy and numeracy in describing and recording patterns of animal behaviour or productivity, and planning management programmes.
4. Stimulation to creative work and encouragement of the aesthetic appreciation of animals.
5. Identification and investigation of the normal range of environmental factors influencing animals leading to an appreciation of the importance of protecting the environment.
6. Contributions to the personal development of students by shared responsibility for animal welfare, establishment of human-animal bonds and caring attitudes, and introduction to potential out-of-school interests involving animals, such as bird watching.

The use of animals in education is sometimes questioned and yet contact with animals is a means not only of learning about behaviour and physiology, but also provides a basis for learning a respect for life and a healthy concern for the well-being of all animals. Attitudes towards animals depend to a great extent on experience in the formative years, much of which will be gained at school. The attitudes displayed by teachers towards animals are likely to be influential, being equally unbalanced if thoughtless and uncaring or unrealistically over-sentimental. Three stages of development in the attitude of children towards animals have been identified: exploitative, emotional and ethical (Paterson, 1990). Very young children tend to be exposed to pets, anthropomorphic literature and television programmes, developing an emotional involvement and sensitivity to animals early. By 14 or 15 they are usually much more objective and less 'sentimental' (Paterson, 1990). Adequate experience of animals in education is therefore important as is an appreciation of how ready and receptive children are at different ages to different learning experiences. Yet recent research has suggested that about 50% of 14-15 year old pupils have no experience of working with living animals in their secondary schools (Lock and Millett, 1992).

The psychologists Zahn-Waxler, Hollenbeck and Radke-Yarrow (1985) have suggested that basic humane attitudes and behaviours are laid down in the first years of life and that teaching a sense of responsibility for animals as well as for other people should begin as early as possible, preferably from a child's first birthday onwards. The use of animals in primary schools needs, therefore, to be carefully thought out so as to be sure that the experience really will enhance children's knowledge and respect for the animals in their care.

The manner and context in which animals are used in education is of overriding importance. If animals are to be used, it is essential that students see them being treated in a sensitive and humane manner, and to good purpose. Balanced discussion of the moral issues surrounding the various human uses of animals should be part of the biology curriculum in schools (Smith, 1992).

Many of us are becoming increasingly removed from the natural world. We live in towns and cities where our direct interactions with nature are all but non-existent. Many enjoy 'domesticated nature' in the form of pets and gardening but still more rate their experiences of 'wild nature' simply for its irritation value: the nettles amongst the roses and the ants in the kitchen. Our ignorance may even breed fear of wild things. Images of animals are frequently sanitised and anthropomorphised. Television programmes show exotic adventures to the Galapagos Islands and the Indian Ocean and the American writer, Joseph McInerney, has described these attempted reconnections to nature, often stylised and frequently glorious, emphasising the stunning diversity and beauty of life on Earth, as parades of charismatic megafauna - beautiful beasts that compete for our compassion on nature shows. It is hardly surprising that the local park and school wormery have difficulty living up to these images. Despite this, it is important to remember that personal contact with living biology - however small and unglamorous - is always exciting.

Rachel Gravett

REFERENCES

ASE, IOB and UFAW (1986) The place of animals in education. *Biologist,* **33**(5), pp 275 - 278.

Department for Education (1995) *The National Curriculum Orders - Science* (England and Wales). London: HMSO.

Lock R. and Millett K. (1992) Using animals in education and research - student experience, knowledge and implications for teaching in the National Science Curriculum. *School Science Review,* **74**(266), pp 115 - 123.

McInerney J. D. (1993) Animals in education: are we prisoners of false sentiment? *The American Biology Teacher,* **55**(5), pp 276 - 280.

Paterson D. (1990) Beastly images of childhood. *New Scientist,* 24 March 1990, pp 53 - 55.

Reiss M. and Beaney N. J. (1992) Living organisms in schools. *Journal of Biological Education,* **26**(1), pp 63 - 66.

Smith J. A. (1992) Dissecting values in the classroom. *New Scientist,* 9 May 1992, pp 31 - 35.

Zahn-Waxler C., Hollenbeck B. and Radke-Yarrow M. (1985) The origins of empathy and altruism. In: *Advances in animal welfare science* (vol. 2, 1986) (ed. Fox M. W. and Mickley L. D.). Dordrecht: Nijhoff (Martinus). pp 21 - 41.

Chapter 2 MICROBIOLOGY

J. Schollar and J. Grainger

ABSTRACT

Although practical microbiology adds a valuable dimension to the study of biology in schools and colleges, the opportunities are not being fully taken up. This is largely because of a lack of familiarity with the topic on the part of teachers and apprehension about safety issues. Microbes - "our invisible allies" - provide benefits to society far in excess of the harm that they cause; indeed, microbes are the base for and sustain all other life. Many suitable and worthwhile investigations can be done at primary and secondary levels to harness pupils' natural fascination with the microbial world, thereby enriching the theory through a better understanding of the diversity of the activities of microbes. Furthermore, insights to the contributions of microbiology to the rapid advances in biotechnology will enable these future citizens to be better prepared when called upon to make informed decisions on controversial issues that will affect society. The purpose of this chapter, therefore, is to provide some encouragement to the teaching of practical microbiology by removing some of the mystery and unwarranted fears that stand in the way of progress in developing the teaching of this fascinating, exciting and important area of biology.

Microbiology is a living science. Having been first recognised as a discipline in its own right some 100 years ago, in more recent times microbiology has, for example, played a central role in the key advances in molecular biology. Now there is the possibility of microbiology becoming central to a major reconstruction of the whole of biological thought if the current tendency continues for research to focus increasingly towards an evolutionary dimension. This prospect has become possible by the realisation some 15 years ago that some bacteria are sufficiently different from the prokaryotes to be reallocated to a new domain, the archae bacteria, establishing two groups of prokaryotes no more related to each other than is either to the eukaryotes. This in its turn breaks down the great and puzzling divide between eukaryotes and prokaryotes.

WHY MICROBIOLOGY?

Microbiology is a fascinating and rewarding topic for the school curriculum. It is unfortunate, however, that its potential is not being fully realised. This is largely because of an understandable lack of familiarity and therefore confidence on the part of many teachers, which stems mainly from the common misapprehension that all microorganisms are harmful and therefore dangerous to use in the school laboratory. While it is undeniable that microbial infections may at best be unpleasant, and at worst fatal, the majority of microorganisms are beneficial, even essential, to our way of life through their activities in, for example, the production of food, medicines and in environmental protection. It is from such activities that examples can be drawn which may be safely used in primary and secondary schools provided that good laboratory practice is observed.

Although microbiology has been known as a discipline for little more than a century, the activities of these minute forms of life had been unknowingly harnessed since

Chapter 2 Microbiology

prehistoric times to produce biochemical changes in materials which were cultivated or collected. These ancient civilisations practised their crafts to produce food and drink by leavening bread, souring milk, maintaining soil fertility, and making beer and wine using principles that can be recognised in some contemporary technologies. Nowadays, the biotechnology industries are able either to select or genetically modify microbes or otherwise improve their performance to produce economically and socially desirable products which meet the growing needs of society.

Biotechnology - a definition
The application of scientific and engineering principles
of materials by biological agents to provide goods and services.

Developments in biotechnology increasingly have a direct effect on our daily life and thereby also increase the range of activities that are appropriate for theoretical and practical study in the classroom. Since many of the activities relate directly to pupils' own experiences (health, disease, food and drink), they have an intrinsic interest in this subject and can easily see the relevance of the work.

Biotechnology - a multidisciplinary activity

microbiology, biochemistry genetics, chemical and biochemical engineering	microorganisms, animal and plant cells, enzymes, antibodies
PRINCIPLES	AGENTS
GOODS	SERVICES
goods, beverages, chemicals energy, pharmaceuticals, pesticides, metal extraction	water purification, waste management, pollution control

---

Biotechnology is a multidisciplinary activity, i.e. it is neither the exclusive domain of one aspect of science, nor of one branch of engineering. Furthermore, it cannot be separated from the social and economic considerations accompanying any major scientific and technological development.

Biotechnology involves the use of biological agents, among which microbes and their enzymes are pre-eminent, to produce goods and services, to the benefit of society. Biotechnology helps to give microbes 'a good press' because it provides examples of their beneficial activities to weigh against the detrimental ones.

Since microbes are of such social and economic importance, it is desirable that an understanding of their role in life should form a significant part of every young person's education. With the development of biotechnology it is even more important that pupils have a full appreciation of the issues raised by this technology.

It is usual for pupils to have a natural interest in microbes, mainly for the following reasons:

1. Although of microscopic size (see Box 1·1), microbes can have profound effects on the largest of animate and inanimate objects.
2. Microbes can grow rapidly and produce growth visible to the unaided eye on natural and man-made substrates.
3. Microbes include a wide variety of different groups with a diversity of properties and activities (see Box 1·2).
4. Some microbes are harmful but many are beneficial.
5. Microbes are ubiquitous; they can be found in virtually all environments including, for example, the extremes of ocean depths and hot springs.

In summary, microbiology is an area of study which generates much pupil interest and offers rewarding opportunities for teaching (see Box 1·3). The benefits gained by pupils from practical work with microbes outweigh the risks supposedly associated with working with them.

The laboratory study of living microorganisms required the development of two major practical methods that are the hallmark of microbiology, i.e. microscopy and aseptic (a much more appropriate word than "sterile") technique. The original purpose of aseptic technique was to protect experiments from contamination from external sources and the understanding that led to its introduction was central to the disproving of the theory of spontaneous generation. In more recent times, we have come to depend on aseptic technique for the containment of microorganisms, thereby protecting the community and the environment from their unwanted effects.

Approximate sizes of the major groups of microorganisms

0.1 mm	=	100 μm	Protozoa, Algae, Moulds
0.01 mm	=	10 μm	Yeasts
0.001 mm	=	1.0 μm	Bacteria
0.0001 mm	=	0.1 μm	Viruses

horizontal bars represent relative sizes
mm = millimetre μm = micrometre

Box 1·1

Major groups of microorganisms

Major groups of microorganisms

Protozoa
Protozoa are found in a variety of habitats such as soil, ponds, lakes, rivers and the sea. They are important in sewage treatment and are involved in many food webs. Very few species cause disease.

Algae
Algae are photosynthetic organisms, found in most aquatic environments (rivers, ponds, lakes and the sea) as well as some terrestrial habitats, such as the surfaces of trees. They are of value to the food industry as a source of colouring agents and as the solidifying agent (agar) for culture media.

Fungi (Yeasts and Moulds)
The fungi include single-celled structures such as the common yeasts and multicellular filamentous branching organisms, *i.e.* the moulds. Fungi are found in abundance in soil and the air. Many are important in food, drink and antibiotic production; some cause diseases of plants and others may spoil food. Mould spores can be allergenic.

Cyanobacteria (Blue-green Bacteria)
Cyanobacteria were once known as the blue-green algae but are now known to be prokaryotes. Some are capable of nitrogen-fixation, others are a food speciality, sold dried in health food shops.

Bacteria
The majority of bacteria are beneficial and of immense value to industry and in environmental protection. A number are important pathogens of human beings and other animals. Some cause food spoilage.

Viruses
Viruses can develop only in the cells of host organisms such as animals, plants and bacteria. Some cause well-known diseases, *e.g.* viruses that attack bacteria, known as bacteriophages or 'phages', are of much value in genetics and molecular biology.

Box 1·2

Microbes can be used to demonstrate principles of biology and to model industrial processes, as well as offering opportunities for teaching across the curriculum.

Biological principles	Industrial processes
population dynamics	industrial chemical,
osmosis	fuels, solvents
photosynthesis	industrial catalysts
genetics, molecular biology	(enzymes)
tropisms	health care products,
enzyme activities	*e.g.* vaccines, antibodies
vitamin assays	foods and beverages,
sulphur, nitrogen and	*e.g.* cheese, yoghurt,
carbon cycles	beer, bread, mycoprotein
biodeterioration, spoilage	waste treatment, *e.g.*
disease	sewage, refuse, pollution control
	oil and metal recovery

Box 1·3

MICROBIOLOGY WITH MICROSCOPES

Much of value in microbiology can be achieved without the use of a high power microscope. The examination of cultures with either a basic plate/dissecting (low power, x5 - x10 total magnification) microscope or a good hand lens can reveal useful information about microbial colonies on agar plates or growth on natural materials, *e.g.* cheese. Features such as the shape, size, colour, elevation, translucency and surface structures illustrate diversity of form among the major groups of microorganisms.

Certain everyday foods, except meat (see *Safe handling of microorganisms*), can provide a safe, inexpensive and very readily available source of a diverse range of organisms for microscopic examination. For example, blue cheese contains the fungus *Penicillium*; yogurt contains the bacteria *Lactobacillus* (rods) and *Streptococcus* (cocci); yeast cells can be seen in the material used for home production of bread, beer and wine; kefir, a mildly alcoholic, acidic fermented milk product, contains various bacteria and yeast; samples of green pond water, and murky water from flower vases and hay infusions are rich in protozoa and algae.

The largest of microorganisms, the protozoa, algae and moulds, can readily be viewed with most microscopes that are found in schools and colleges. Magnifications of up to x100 (low power, x10 eyepiece and x10 objective lenses) provide a large field of view and adequate detail with which to recognise them. The use of a high power dry objective lens (x40) reveals more detail but the smaller field of view can be restricting, particularly with motile (moving) organisms.

Protozoa and algae can normally be viewed by placing a drop or several loopfuls of a sample on a slide and covering it with a coverslip, and examining using the low power objective. By including some debris from the sample, the numbers viewed are increased because of the large concentration of organisms at surfaces, particularly protozoa that are grazing on the large numbers of bacteria located there. A semi-permanent preparation to last through a practical lesson without drying out can be made by sealing the edges of the preparation with molten "Vaseline" applied with the narrow side of a warmed microscope slide. Motility can be slowed by including strands of lens tissue or carboxymethyl cellulose solution in the preparation; iodine solution stops movement but kills the cells. Moulds can be transferred by placing a small piece of transparent adhesive tape on the specimen and then transferring the tape on to a slide. Normally, yeasts can also be viewed unstained at these magnifications.

Bacteria can be seen clearly only if the instrument has an oil immersion objective lens and the specimen is illuminated by a lamp. The lens has a magnifying power of usually x100, providing a magnification of x1000. The front lens of the objective is brought into contact with a drop of special oil (oil immersion) which has been placed directly on the stained bacterial specimen; a coverslip is not used. Since there is no air gap and immersion oil has the same refractive index as glass, the light passing up through the microscope slide, oil and objective lens is not refracted. This produces a clearer and brighter image. Bacteria in a stained, heat-fixed smear (see below) viewed with a x40 objective lens can just be seen but the image is very small and at best only the overall shape and arrangement of the cells can be distinguished.

Bacteria must be stained to be seen clearly. A simple stain with a basic dye, *e.g.* crystal violet, is perfectly adequate, but another widely used procedure is Gram's staining method. Gram, a Danish scientist, discovered that certain bacteria could be distinguished from the tissue of an infected animal. This led to a new staining protocol which revealed two different staining reactions among bacteria: Gram-positive (stain purple) and Gram-negative (stain pink). Differences in biochemical composition of the cell walls of Gram-positive and Gram-negative bacteria account for this difference in reaction. This distinction has provided a useful means for the division of bacteria into two major groups.

MICROBIOLOGY WITHOUT MICROSCOPES

The observation of microorganisms through a microscope provides an invaluable way of understanding microbiology. However, a shortage of suitable microscopes need be no bar to practical microbiology in schools, for there are many activities that do not require them. Indeed it is worth pointing out to a budding Pasteur or Fleming with an aversion to microscopy that professional microbiology by no means has to involve the use of the microscope. Growth of microorganisms and the effects of their growth can be readily seen by the unaided eye on natural materials and in test tubes and Petri dishes. The source of the growth may be a pure culture of a suitable organism obtained from a reputable supplier or some natural material recommended as being appropriate for use in schools.

The environment and food are rich, convenient and educationally valuable sources of microorganisms but one must bear in mind that some can harbour pathogenic organisms. Nevertheless, by choosing samples wisely and using good laboratory practice (see *Safe handling of microorganisms* section below) it is safe to make studies of food, water, soil and air to show the variety and diversity of microbial life. For example, although meat and meat products should not be used because they can be sources of food poisoning organisms, vegetables such as peas pose substantially less risk. Teaching programmes that include investigations of food production and preservation/hygiene and the decay of biological materials lend themselves particularly to practical activities at all levels including primary.

A variety of readily available general purpose media can be used to promote the growth of microorganisms. For example, moulds and yeasts grow well on malt or potato media; nutrient agar (or broth) is a suitable medium for the growth of many bacteria, supplemented with a sugar, commonly glucose, for fermentative types. Indicator media can also be useful. For example, clear zones around colonies grown on milk agar medium show hydrolysis of casein, the milk protein; a similar observation for starch hydrolysis can be made after flooding a starch agar medium with iodine solution; with china blue lactose agar medium it is possible to recognise colonies of lactose-fermenting bacteria that are found in milk and milk products. Selective methods can also be of value, *e.g.* using a medium with an acid pH value (5.5 as with malt agar) to isolate yeasts, or heating a soil suspension (say, at 80°C for 10 minutes) to destroy all but bacteria that possess heat-resistant endospores.

A note of caution must be introduced here concerning some culture media that have been designed for the professional medical or foods microbiology laboratory and should not normally be used for school investigations. Examples include the incorporation of blood in culture media which provides nutrients necessary for the growth of some animal pathogens, and the use of MacConkey agar medium on which colonies of pathogens will dominate because others are inhibited by the bile salts present in the medium. Such media may be recommended in texts that are either out-of-date or published abroad. Another source may be the professional microbiologist who, whilst eager to offer support, is unfamiliar with school facilities. Such invaluable willingness to help should not be spurned, but tactfully harnessed into providing industrial or commercial background, talks, demonstration material or visits.

SAFE HANDLING OF MICROORGANISMS

After having been confronted with the relevant safety guidance it would be most unfortunate, but understandable, if educators unfamiliar with microbiology were to be discouraged from wanting to explore this fascinating and rewarding area of study. However, the requirements of good laboratory practice need not be as fearsome as they may at first seem; indeed, there are practical activities that are appropriate for primary schools as Box 2·4 on appropriate levels on practical work indicates. To keep matters in proportion, appropriate investigations in microbiology are no less safe than many other activities undertaken in science and technology laboratories. Moreover, there are many that would argue that with proper risk assessment evaluations then the laboratory should be one of the safest places in school.

Appropriate levels of practical work

Levels are based on the degree of hazard represented, the type of training required by the teacher to perform the work with pupils, and the facilities available.

Level 1
Observational activities with microorganisms or materials that contain them which have little, if any, known risk.
May be performed by teachers with no specialist training.
The most appropriate organisms are certain yeasts, moulds and algae.
Simple culture work is possible in containers that cannot be opened by pupils.

Level 2
Activities with known microorganisms using suitable media and culture conditions.
Microbial cultures should be obtained from reputable suppliers.
Cultures containing microorganisms obtained from the environment should not be opened by pupils.
Environments likely to contain human pathogens, e.g. lavatories, body surfaces, human fluids, meat, raw milk, must not be used.
Teachers should have a working knowledge of safe practice in aseptic technique and an appreciation of safe disposal techniques which may be gained from attending a short INSET course in microbiology.
The subculturing and staining of bacteria by pupils is inadvisable.
The most appropriate organisms are certain yeasts, moulds, algae and bacteria.

Level 3
Pupils can subculture known bacteria and fungi obtained from reputable suppliers.
Good aseptic technique is required by both staff and students.
This level requires that staff have thorough microbiological training and recent practice.
The most appropriate organisms are certain yeasts, moulds, algae, bacteria and viruses.

Box 2·4

Chapter 2 Microbiology

SAFETY LEGISLATION

It is emphasised in safety legislation that the handling of microscopic organisms presents unique problems. Under the Health and Safety at Work Act (1974) and Control of Substances Hazardous to Health (COSHH) Regulations, educators involved in the teaching of microbiology in schools and colleges are specifically obliged to understand and implement the recommendations made in general risk assessment documents, which include: *Topics in Safety* (ASE, 1988), *Microbiology: an HMI Guide for Schools and Further Education* (DES, 1990), *Laboratory Handbook* (CLEAPSS, 1992) and *Safety in Microbiology (A code of practice for schools and non-advanced further education)* (Committee for Safety in Microbiology, 1991). It should be noted that before one can embark on a practical activity an assessment of risk must be performed and as a result of the outcome of the assessment suitable precautions or actions taken (see chapter on Safety). Official guidance needs to be constantly reviewed and clarified in the light of experience, and developments in techniques and the understanding of microbiology.

LEVELS OF WORK AND GOOD LABORATORY PRACTICE

The nature and level of work should be determined by the experience of the teacher and the facilities (see Box 2·4). In general, the levels of work appropriate to different ages are: age 5-11, Level 1 only; age 11-16. Levels 1 and 2 only; age 16+, Levels 1-3. Some variation is appropriate for vocational courses in further education. Microbiological techniques focus on the development of good laboratory practice (see Box 2·5) which minimises the risks to the operator or others in the vicinity - and should also prevent contamination of cultures.

There are five characteristic aspects of practical microbiology which make thorough planning ahead essential, especially at Levels 2 and 3:

1. Preparation and sterilisation of equipment and culture media.
2. Preparation of cultures.
3. Inoculation of the media.
4. Incubation of cultures and, if appropriate, sampling during growth.
5. Sterilisation and safe disposal of cultures and decontamination of equipment.

GOOD LABORATORY PRACTICE - A SUMMARY

SAFETY PRECAUTIONS

- ➢ Wear protective clothing.
- ➢ Wash hands before and after working with microorganisms.
- ➢ Wipe down work surfaces with a suitable disinfectant.
- ➢ Cover cuts or scratches with a waterproof plaster.
- ➢ No eating, storing of food, drinking or smoking in the laboratory
- ➢ Do not perform hand to mouth operations, e.g. mouth pipetting, licking gummed labels, sucking pens or pencils.
- ➢ Treat all cultures as potentially pathogenic; do not remove cultures from the dedicated area in the laboratory.

Box 2·5

Good Laboratory Practice - a summary (continued)

> Do not make cultures from potential sources of pathogens, *e.g.* blood, skin, mucus, pus from cuts, urine, faeces, meat, raw milk.
> Do not incubate cultures at 37 °C or above because organisms that grow well at human body temperature include some pathogens; 30 °C should be regarded as the upper limit.
> Do not make cultures from samples incubated under anaerobic conditions because organisms that grow in the absence of oxygen include some pathogens.
> Obtain cultures only from reputable suppliers.

Aseptic technique
> Work near a Bunsen burner so that air-borne organisms which might otherwise contaminate the work are carried away by the resultant updraught.
> Flame the necks of bottles/tubes to ensure an updraught which helps prevent contaminants from entering. It should be noted that if the bottle neck is heated too strongly, rapid cooling may draw air into the vessel with resulting contamination.
> Open cultures and sterile equipment for as short a period of time as possible so as to reduce the possibility of contamination.
> Correctly flame sterilise loops and wires before and after use to prevent the contamination of the operator, environment or culture.
> All equipment and media to be used for microbiology should be autoclaved or sterilised chemically before and after use.

Box 2·5 (continued)

SUITABLE CULTURES

The practice of treating all cultures as potential pathogens forms the basis of good laboratory practice. The organisms on approved lists are drawn from science teaching projects which present minimum risk. The lists cover the usual needs of school microbiology. This does not mean, however, that other organisms not hitherto used are necessarily unsuitable. Teachers having a specific reason for wishing to use other organisms should seek professional advice from for example the Microbiology in Schools Advisory Committee (MISAC), the National Centre for Biotechnology Education (NCBE) which works closely with MISAC or the CLEAPSS School Science Service. An example of an extension to the list is the acceptability of the bacterium *Leuconostoc mesenteroides*, a close relative of two of the listed organisms, *Lactobacillus* sp. and *Streptococcus lactis*. On the other hand, advances in isolation and culture techniques sometimes reveal new examples of unusual strains of traditionally 'safe' species that have become associated with disease. However, this should not be taken as the signal for the automatic banning of the entire genus or species, rather as an indication to seek professional advice.

Chapter 2 Microbiology

ARE THE BENEFITS WORTH THE TROUBLE?

In addition to learning from the procedures and outcome of practical activities in microbiology, there is much to learn about microorganisms and their activities from understanding the reason for the various aspects of good practice - and what better than a sensitivity of students to the "hidden" hazards of microbiology to encourage a sense of respect and responsibility for the well being of others?

"What is it?" This is a frequent and understandable question for a student to pose and reflects an interest that should be nurtured. However, the identification and classification of a particular microorganism with precision is of limited educational value in the school context and anyway is normally beyond the resources of a school or college laboratory. To identify microorganisms fully is a specialised task that is based on the examination of many biochemical, physiological and, increasingly, genetical features as well as morphological ones - so specialised indeed that it would be a brave brewery microbiologist who would try to identify an organism isolated from a throat swab in a pathology laboratory! It is perfectly adequate for present purposes to limit an identification to one of the major groups (see Box 2·2), possibly also suggesting a likely genus if it is a well known one in the material under study, *e.g.* the yeast *Saccharomyces* in beer. It is of much greater value to spend time considering how and why the organism is growing under the prevailing conditions and what might be the consequences.

WHAT WE SHOULD KNOW BEFORE WE START

VESSELS AND CONTAINERS

When starting on microbiology the first requisite is an adequate supply of containers for the preparation and storage of media and for the growth of microorganisms. The following selection of materials are useful for a variety of microbiological investigations.

McCartney bottle (narrow neck); **Universal bottle** (wide neck)

These are glass bottles of 30 cm^3 capacity with an autoclave-plastic or metal screw cap. When used for microbiology and the production of sterile media they should be not more than half full, *i.e.* 15 cm^3. These bottles, when containing a slope of agar medium (5 cm^3 only), are very valuable for maintaining stock cultures. Stocks of agar and broth media can be prepared, sterilised and stored (in the dark) for several weeks before they are required. Caps should be slightly loosened for culturing strictly aerobic organisms but re-tightened before use in the class laboratory. A few minutes of heating a bottle of agar medium at about 100°C will melt the medium which can then be poured aseptically into a Petri dish to produce an agar plate.

Medical flat (flat bottle with parallel sides); **'Duran'** (cylindrical bottle)

These bottles range in size from 50 cm^3 to more than 1000 cm^3 and are closed with an autoclavable plastic or metal screw cap. They are very valuable for production and storage of media and sterile water. It is important that they are not over-filled; they should not be more than two thirds full to ensure that the medium does not boil over during autoclaving. Medical flats of 250 cm^3 capacity are convenient for preparing, sterilising and storing agar media to provide a stock that can be melted to produce agar plates when needed.

Conical flask (most useful size: 250 cm^3)

These are very convenient for growing microbes in liquid culture. A large flask filled with a small volume of broth provides a large surface area-to-volume ratio providing aerobic organisms with the good aeration necessary for growth. Ideally only 100 cm^3 of broth is used in a 250 cm^3 flask. The neck of the flask should be correctly plugged with non-absorbent cotton wool and during sterilisation protected from the steam in the autoclave by covering it with either aluminium foil or greaseproof paper held in position with two elastic bands. A flask forms the basis of the simplest type of fermenter that can be constructed.

Test tube (15 mm diameter)

Test tubes can be used for the production of agar slope cultures and broth cultures. Normally only 5 cm^3 of medium is used and the tubes are closed with non-absorbent cotton wool plugs or plastic or foil caps. They are not always as convenient as Universal or McCartney bottles since test tubes containing media cannot be stored for long periods of time at room temperature owing to evaporation of the contents; also racks are needed to support them for use in the laboratory. (See under **Conical flask** for keeping plugs dry during autoclaving.) They are also much more fragile than McCartney or Universal bottles.

Petri dish (most common size: 90 mm diameter).

Traditionally Petri dishes were made of glass, which allowed re-use after having been autoclaved, washed, cleaned and re-sterilised. It is now more usual to use plastic disposable Petri dishes which are supplied in sterile packs. They are used only once for growing cultures and are then chemically disinfected or, preferably, sterilised by autoclaving and discarded with their sterile contents to the dustbin or incinerator. Most Petri dishes have three small lugs in the lid that raise it less than a millimetre above the base to assist gas diffusion, *i.e.* oxygen in and carbon dioxide out, but the overhang of the lid prevents unwanted organisms from entering provided that a film of liquid does not breach the gap. Plate production involves aseptically pouring molten, cooled (*c.a.* 45-50 °C) medium into a sterile Petri dish and then the solidification of the medium. The inoculated medium is inverted for incubation "upside down", *i.e.* the medium-containing half uppermost, to ensure that condensation which may form on the lid cannot drop onto the agar and diminish opportunities for contamination.

Tools of the technique

The purpose of the discipline of aseptic technique is to prevent both the contamination of the environment and worker by the microbes being handled, and contamination of a pure culture by microorganisms from the environment. It should be noted that aseptic procedures should be carried out as quickly as is safely possible to minimise the risk of contamination. During the transfer of cultures all containers should remain open for as short a time as possible and open necks of vessels inclined towards and close to a semi-roaring Bunsen burner flame. A number of instruments are required for the aseptic transfer of cultures of microorganisms.

Wire loops are used to transfer microbes from one container to another. They are suitable for use with agar and broth cultures. The instrument consists of a handle which holds a replacement nichrome or platinum wire which is formed into a small loop. The loop is sterilised before and after use by heating to redness; this procedure is known as "flaming" and is a quick and effective sterilisation technique.

The loop is held in the outer part of the Bunsen burner flame (semi-roaring) pointing almost vertically downwards and kept there until the whole length of the wire becomes red hot. The loop should be cooled by holding it close to the flame before use.

The loop must not be used while it is still hot as the heat would kill the organisms and also, more seriously, it might cause spluttering and release organisms to the air. If the loop produces an audible hiss when touching the medium, it is too hot. If the operator wishes to check that the loop is sufficiently cool, it can be run down the inside glass of the culture bottle or touched on an area of agar in a Petri dish which has no microbial growth.

A used (contaminated) loop must not be placed on the bench; *immediately* after use it must be flamed again, this time being brought down into the flame very slowly so that any liquid or organic material is vaporised. If a liquid-filled loop is brought down too quickly droplets could be forcibly ejected from the flame and contaminate the surrounding area or the operator.

Straight wires are constructed from the same material as loops and are used for transferring from agar cultures. Wires are sterilised in the manner similar to that used for loops. Straight wires can also be used for "picking off" individual colonies from agar plates. If colonies are very close together, use of a straight wire will permit transfer from a single, separate colony much more reliably than by a loop.

Spreaders are made of glass bent into an 'L' shape and used to spread organisms evenly over the surface of an agar plate. Spreaders can be sterilised by wrapping in foil or greaseproof paper and autoclaving. They must not be sterilised by heating in the flame of a Bunsen burner because of the danger of the glass shattering.

Spreaders can be disinfected by placing in 70% (v/v) alcohol and passing through a flame to ignite the alcohol remaining on the head, thereby raising the temperature but not sufficiently to break the glass. *The head must be directed downwards while the alcohol burns away* and then the head is allowed to cool before use. After use the contaminated spreader is placed back in the alcohol and stored there until needed again. *The operator must not put a flaming spreader in the alcohol and care must be taken at all times to prevent accidents with burning alcohol.*

Inoculating chambers (transfer chambers) are designed to minimise the risk of contamination by airborne organisms and to provide an area of containment. However, they are not essential for school or college work and in some cases will create problems. For example, a laminar flow cabinet can be a very valuable piece of equipment for maintaining sterility whilst dispensing culture media aseptically but not for inoculation purposes when the direction of the air flow blows microbes towards the operator.

Pipettes are used for the transfer of sterile liquids and cultures in broth media. They are made of either glass or plastic, graduated or calibrated to deliver a measured volume with a degree of precision appropriate to the application, *e.g.* 10 cm^3 or a specified fraction of 1 cm^3. The name "Pasteur pipette" or "dropping pipette" is commonly given to the type which is designed to deliver fluid drop-wise from an extended, narrow tip drawn from one end of a short length of c 5 mm diameter glass or plastic tubing. The volume of the drop is either unspecified or known. In the latter case the volume is determined by several factors including the external diameter of the tip, the rate of drop delivery and the inclination of the pipette which should be held vertically; 0.02 cm^3 is a common drop volume (hence the term "50-dropper"). A device for delivering measured volumes from a non-calibrated pipette can also be constructed by using a short length of tubing to connect, for example, a Pasteur pipette to the barrel of a syringe in place of the needle.

Prevention of contamination of sterile liquid or culture from the surrounding environment during transfer and protection of the operator and environment by containment of microorganisms within the equipment are achieved by the inclusion of a short plug of non-absorbent cotton wool in the neck of the sterile pipette. Pipettes are either supplied in sterile packs or packed in plugged test tubes and wrapped in greaseproof paper and sterilised in the laboratory. Sterilisation is achieved by autoclaving or, for glass pipettes only, hot air.

A rubber teat is an inexpensive and convenient device for drawing fluid into and expelling it from a pipette - *pipetting by mouth is not permitted*. It is essential that the cotton wool plug in the neck of the pipette remains dry throughout the sterilisation procedure and transfer operations otherwise it will not function as a barrier to the passage of microorganisms. If improperly used with a microbial culture, a pipette can cause a hazard in the work area through the release of aerosols, *i.e.* droplets of liquid, which contain microorganisms. This hazard is avoided by careful teat control which involves gradual changes in pressure, expulsion of fluid only within the confines of the vessels, and avoidance of taking in air when charging a pipette. It is prudent to perfect the technique by practice with water beforehand.

A used (contaminated) pipette must not be placed on the bench; *immediately* after use it must be immersed totally in disinfectant such that the entire inner surface is decontaminated. The teat must remain in position until the pipette is over the pot of disinfectant, otherwise drops of residual liquid could fall from the pipette and contaminate the working area.

CULTURE MEDIA AND INOCULATION PROCEDURES

An **agar plate** is a layer of solidified neutral medium in the bottom of a Petri dish on to the surface of which microbes are inoculated. With suitable incubation the microbes develop into colonies that can be clearly seen with the naked eye. Agar is a complex polysaccharide that is extracted from seaweed (*Laminaria hypoborea*) and is the solidifying agent added to a solution of nutrients. The nutrients that are incorporated are varied according to the needs of the microorganisms to be cultured, *e.g.* nutrient agar, commonly containing enzymatic extracts of meat and possibly yeast, will support the growth of a range of bacteria but requires supplementation with glucose for fermentative types; mannitol yeast agar medium is particularly suitable for the growth of the bacterium *Rhizobium*. Cultures that have been kept for some time on an agar slope medium in a refrigerator require subculturing more than once before they are sufficiently active for reliable use in the classroom.

A **streak plate** is made by sequentially diluting an inoculum over the surface of an agar plate using a wire loop (see Figure 2·1). Streak plates are very useful for separating mixed cultures so that an isolated colony can be subcultured on to fresh medium for subsequent examination. Streak plates can also be used to check the purity of a culture. Contamination is revealed when two or more different colony types are seen. It is good practice when subculturing a stock culture to check that it is still a pure culture. It may be assumed that when only one type of colony can be seen on the plate the culture is pure. However, this is not always necessarily so and a microscopical examination may be necessary because different organisms may have a similar colony form - but it is a good guide for present purposes.

A **pour plate** is produced by the mixing of a diluted culture of microbes with about 15 cm^3 of molten agar medium. Microbes are dispersed throughout the medium and, after cooling to solidify the medium and incubation, colonies can be seen throughout the medium. There is a danger that if the molten medium is kept too hot, the heat may kill many of the organisms; but if the medium is allowed to cool too much it may solidify prematurely. The medium cannot be remelted without raising it to about 100 °C, a temperature that will kill most microorganisms. Therefore it is important to keep the agar medium molten in a water bath between 45 °C and 50 °C. Appropriate dilutions of a culture or sample will yield separate colonies after incubation, thereby providing quantitative data.

A **lawn plate** is one on which a culture of microbes is spread over an already solidified plate of agar medium to produce a confluent lawn of growth after incubation. This can be achieved by aseptically adding a small volume of broth culture (say 0.1 cm^3) to the surface of the agar medium and then using a sterile glass spreader to distribute the culture evenly over the surface of the medium. Alternatively the culture may be spread by gently tilting the plate.

A **spread plate** is similar to a lawn plate except that the inoculum is a culture or sample that has been diluted to provide separate colonies rather than a lawn, thereby providing quantitative data.

An **agar slope** is made by adding molten sterile agar medium to either a test tube or bottle and allowing the medium to solidify with the vessel in a sloped position. This provides a wedge of medium with a large surface over which to streak an inoculum. Agar slopes are commonly used for providing pure cultures for class use and for maintaining stock cultures.

Broth media are liquid media that contains exactly the same nutrients as the equivalent agar medium but without the solidifying agent. The broth is best distributed in screw cap bottles which are more convenient than test tubes for reasons of stability and long storage.

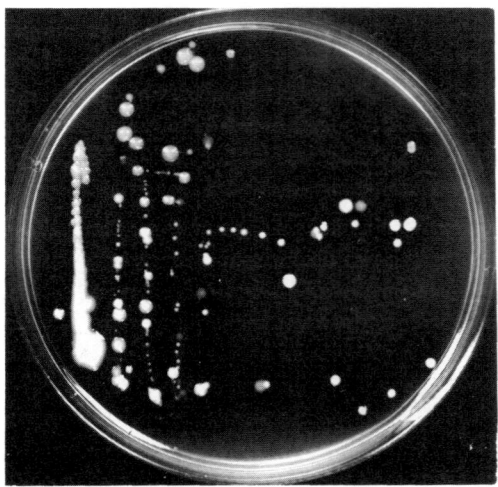

Figure 2·1. A streak plate.

"KILLS ALL KNOWN MICROBES - DEAD!"

STERILISATION

Sterilisation is the complete destruction or removal of all forms of living organisms. Once a material is sterilised it will remain sterile indefinitely if properly sealed (*cf.* Pasteur's swan-necked flask equipment). There is no such state as being "nearly sterile". The (endo)spores of certain bacteria are the forms most resistant to heat and chemicals. Of the various methods available for achieving sterilisation heat is usually the one of choice for convenience and effectiveness.

Dry heat:

Flaming is often used for sterilisation of inoculating loops and straight wires. **Incineration** is an effective way of destroying infected material or equipment that is not needed again. **Hot air** destroys microorganisms by oxidising their cellular components. This requires a high temperature (160 °C) and, since air is not a good conductor of heat, a long exposure time (1 or 2 hours) is required.

Heat in the presence of moisture:

Heating at 100 °C (boiling water) does not ensure sterilisation because some bacterial endospores can survive exposure to this temperature for considerable periods. The most common method of achieving sterilisation in the presence of moisture involves the use of an autoclave, normally operating at 121 °C, a temperature only achieved in *pure* steam (*i.e.* with no air present) at a pressure of 103 kPa (kilo Pascals) or 15 pounds per square inch. It should be noted that at this pressure a steam-air mixture does not achieve that temperature and thus does not guarantee sterilisation. Sterilisation time consists of three components: *penetration time* - the time taken for the least accessible part of the load to reach the required temperature; *holding time* - the minimum time in which, at a given temperature, all living things will be destroyed as a result of hydrolytic denaturation of cell components; *safety time* - the safety margin, usually half the holding time. For normal school use with a normal load the total sterilisation time would be 20 minutes, including about 5 minutes of holding time. For small loads and empty equipment the penetration time would be negligible, whereas a large volume of medium would require a much longer penetration time. Some autoclaves now operate at 126 °C which reduces the sterilisation time.

Pasteurisation is a process that uses less extreme heating conditions to reduce the microbial population, as in pasteurised milk, but it does not achieve sterilisation.

Filtration:

Liquids that are heat-sensitive can be sterilised by passage through a sterile filter pad connected to a sterile connecting vessel. Removal of microorganisms is achieved largely by adsorption, not by mechanical retention. In consequence, particles smaller than the pore size of the filter are retained. The used pad is sterilised and discarded.

Radiation:

Gamma rays are commonly used commercially for the sterilisation of plastic items, such as syringes and tissue culture vessels. Gamma rays are ionising radiations, a powerful form of radiation. They cause the formation of free radicals which react with and inactivate macromolecules including DNA, leading to the death of the microbial cells.

Chemical disinfection

Disinfectants are chemical agents which destroy microorganisms but not, ordinarily, bacterial endospores, *i.e.* they cannot be relied upon for *sterilisation*. "Disinfectant" is usually used in relation to inanimate objects, whereas "antiseptic" refers to contact with the body.

Ethyl alcohol (70%, v/v, solution) has good "wetting" (detergent) properties and is effective for cleaning equipment. The mode of action includes protein denaturation which is enhanced with increase in temperature.

Domestos (5%, v/v) and Chloros (1% v/v) are chlorine-based (hypochlorite) disinfectants commonly used in discard pots on the bench for receiving pipettes immediately after use. Their mode of action is through the release of chlorine which acts as an oxidising agent, but their effectiveness is reduced by the presence of organic matter. Solutions should be used soon after preparation because a high concentration of free chlorine must be maintained for hypochlorite disinfectants to be effective. Care should be taken when handling the concentrate and suitable clothing and eye protection should be used in the make up of classroom dilutions of the disinfectant.

Hycolin (2%, v/v) is a clear soluble phenolic disinfectant which contains both a green pigment and a detergent to aid its action which is protein denaturation. This category of disinfectant has a broad spectrum of activity but is only weakly active against bacterial endospores. There is little inactivation by organic matter and it is commonly used in the laboratory to treat accidentally spilled cultures.

The diluted solution is very stable and will maintain its antimicrobial activity for at least a month. Undiluted disinfectant should be regarded as "irritant" and care should be taken to avoid contact with the skin and eyes; thus it is recommended that to avoid accidental contact with the body suitable protective clothing should be worn and eye protection when dispensing the undiluted disinfectant.

Tegodyne is an iodophor, a solution of an organic compound of iodine. The released iodine, like chlorine, combines indiscriminately with organic matter. Proprietary products, such as Tegodyne, are multi-purpose disinfectants suitable for wiping over the work area before and after practical work. Again like the above proprietary disinfectant, eye protection and plastic or rubber gloves are advised when handling undiluted material.

Educational suppliers provide a range of these types of disinfectants for specific uses under various proprietary names and purchasers should select the most appropriate for the job it has to do.

PRACTICAL INVESTIGATIONS AND PROJECT WORK: FOR YOUNGER PUPILS.

E.G. KEY STAGES 1 AND 2

Investigation of the effects of disinfectants and sterilising agents using yeast

Aspects of health and hygiene can be illustrated by investigating the effect of commercial disinfectants and sterilising solutions on yeast, a suitable microbe for this level of work. A yeast slurry with a little more added sugar (sucrose) will ferment and produce a frothy head, the extent of which in a tall, narrow container can be used to assess the effectiveness of various concentrations of disinfectants and sterilising solutions. The investigation can be extended to demonstrate the action of detergents by assessing the amount of froth in the presence and absence of a detergent.

Investigation of decay/deterioration of materials

A valuable initial problem to put to pupils is the assembly of a safe container in which to examine decay. The container must be transparent to allow for the observation of decay and allow gases to escape yet retain any microbes and their spores within the vessel. One convenient approach is to use an empty jam jar with a lid in which there is a small hole closed with a plug of non-absorbent cotton wool; the lid should be secured in place with adhesive tape to prevent pupils opening the jar. Animals and animal products (meat or gravy) must not be used. After the investigation the container should be carefully opened in disinfectant.

Investigation of the role of yeast in bread production

Bread making is one of the oldest uses of yeast. The yeast produces carbon dioxide which causes the dough to rise and small amounts of alcohol which evaporate when the bread is cooked. Practical activities can be devised to investigate the effects on dough rising of different strains of yeast, types of flour, temperatures and additives such as vitamin C and the enzyme -amylase. It is easy to see and measure the rate of dough rise in a tall, narrow container such as a measuring cylinder.

Production of real ginger beer using yeast, fresh ginger and sugar

Different yeast strains are used for the production of different alcoholic drinks, *e.g.* beers as distinct from lagers. Ginger beer is a very mildly alcoholic drink that can be made in seven days and can be used to demonstrate the principles of fermentation. Many popular cookery books have recipes for the production of ginger beer using freshly chopped ginger. Plastic bottles are excellent for the production and storage of the beer, but sealed glass vessels must not be used because they may explode under pressure from the gas generated by fermentation.

Food production

Many dairy products are produced with the help of microbes. Yogurt is a good example which lends itself to investigations in the classroom. Cheeses, especially blue cheeses and those with a white crust, are valuable for observational work; the fungus *Penicillium* can be seen in the blue veins and in the crust. Other products such as Quorn, a novel food consisting of a fungus (which is why it is known as "mycoprotein") can be used for discussions of a number of aspects of food technology and nutrition.

Growth of Oyster cap mushrooms on toilet rolls

Supermarket shelves are often stocked with various types of speciality mushrooms. The oyster cap mushroom is becoming increasingly popular. It is produced commercially on waste cellulosic material or tree stumps. In the classroom it can be grown on moist toilet roll, a convenient alternative to a tree! The oyster cap starter culture can be obtained from certain garden centres or NCBE.

Food preservation

Methods for the preservation of foods can be easily and safely examined by the use of frozen vegetables, *e.g.* peas. It is possible to investigate the effects of different preservative solutions (vinegar, salt, sugar) or exposing them to various treatment (drying, warming, boiling, storing) on the spoilage of peas. After a day or so at room temperature when spoilage has taken place, the liquid containing the peas becomes turbid. If the liquid remains clear, little or no microbial growth has occurred and the peas have been preserved. Such peas should never be eaten, no matter how well the preservative worked! Frozen peas should be used for this type of investigation because they present minimal

Chapter 2 Microbiology

risk from the potential danger of growing harmful food spoilage organisms. Meat and meat products are likely to contain large numbers of such dangerous organisms, and must never be used for this type of investigation.

Investigations of microbial enzymes in biological washing powders

Enzymes from microbes are included in biological washing powders to assist in the cleaning of soiled clothes. Practical activities investigating the action of the enzymes can be readily carried out in the classroom. The action of amylase, which degrades starch, can be examined by seeing how well biological and non-biological powders remove ironing starch spray from cloth samples. The action of protease, which degrades protein, can be examined by its effect on gelatin (a protein) which holds the silver halide crystals on exposed and developed photographic negatives. Proteolytic enzymes degrade the gelatin and cause the gelatin layer to fall off the plastic backing, releasing the crystals.

PRACTICAL INVESTIGATIONS AND PROJECT WORK: FOR OLDER PUPILS

(E.G. KEY STAGES 3 AND 4, AND POST-16)

Physiology and growth
- Comparison of the relative sizes of bacteria, yeasts and algae using a microscope eye piece graticule.
- Sorting bacteria (*Micrococcus, Escherichia* and *Bacillus*) by simple biochemical tests.
- Effect of heat on the survival of soil microbes.
- Factors affecting the generation of light by the bacterium *Photobacterium phosphoreum*.
- Factors affecting the growth of the marine bacterium *Vibrio natriegens*.
- Evaluation of growth rates of fungi, *e.g. Mucor, Neurospora*, on a different agar media.
- Fermentation of different sugars by the yeast *Saccharomyces cerevisiae*.
- Comparison of different sugar fermentations by the yeasts *S. cerevisiae, S. diastaticus* and *Kluyveromyces lactis*.
- Investigations on the action of the yeast *Saccharomyces diastaticus* on starch-containing media.
- Action of yeasts and bacteria in the generation of electricity in a microbial fuel cell.
- Production of poly-ß-hydroxybutyrate, a basis of biodegradable plastic, by the bacterium *Alcaligenes eutrophus*.
- Entrapment of yeast to demonstrate the principles of immobilisation of microbial cells and enzymes.
- Formation of protoplasts with microbial enzymes.
- Sensitivity of the alga *Euglena* to light of different wavelengths.
- Sensitivity of the protozoan *Paramecium* to light of different intensities.

Control of microbial growth
- Evaluation of the effectiveness of disinfectants on microbes such as the yeast *Saccharomyces cerevisiae*.
- Effects of toothpastes and mouthwashes on the bacterium *Micrococcus luteus*.
- Effects of deodorants, antiperspirants and skin cleansers on microbes.

- Effects of washing up liquids on microbes.
- Effects of antiseptics, *e.g.* TCP, on microbes using gradient plates.
- Sensitivity of microbes to antibiotic using diffusion methods
- Action of nisin, an antibiotic, on microbes.

Food
- Examination of the growth of the fungus *Penicillium* in blue cheese.
- Production of sauerkraut using the natural bacterial flora of cabbage.
- Production of tempe, an Indonesian fermented food.
- Production of yogurt.
- Investigation of the action of penicillin on yogurt production.
- Investigation of the action of bacterial starter cultures in yogurt production.
- Production of ß-carotene, a colouring agent, by the alga *Dunaliella*.
- Role of the yeast *Saccharomyces cerevisiae* in bread-making.
- Investigations of the growth of microbes that colonise bread.

Beverages
- Production of wine or cider vinegar by the bacterium *Acetobacter aceti*.
- Production of ginger beer using the yeast *Saccharomyces cerevisiae*.
- Tolerance of different strains of wine and beer yeasts to alcohol.
- Comparison of action of the yeasts *S. cerevisiae* and *S. carlbergensis* in the production of beer and lager.
- Comparison of different types and numbers of microbes from bottled water (carbonated) and still (mineral and spring).

Agriculture
- Investigations of the nitrogen cycle, *e.g.* isolation of the nitrogen-fixing bacterium *Rhizobium* from root nodules of leguminous plants such as clover; enrichment of the nitrifying bacterium *Nitrosomonas* in soil.
- Making silage and the effects of additive that contain microbes, *e.g.* the bacterium *Lactobacillus plantarum*, or cellulolytic or pectinolytic enzymes.
- Demonstration of DNA transfer by the action of the bacterium *Agrobacterium rhizogenes*.

Industrial enzymes
- Production of extracellular cellulases by the bacterium *Cellulomonas*.
- Production of extracellular lipase by the yeast *Saccharomyces cerevisiae*.
- Production of extracellular amylase enzyme by the fungus *Aspergillus oryzae*.

Environment/ecology
- Comparison of the microbial flora of different aquatic habitats.
- Isolation of fungi and bacteria from soil and the surfaces of plant leaves.
- Action of the fungus *Trichoderma* in the degradation of cellulosic material.
- Action of the bacterium *Thiobacillus* in the microbial leaching of metals.

Check List and helpful hints

Obtain 'Topics in Safety' (ASE, 1988) and 'Microbiology and HMI guide' (DES, 1990)
Perform assessment of risk for proposed practical investigations.

Obtain an autoclave or pressure cooker for sterilising media and equipment
Only use distilled or demineralised water in the vessel to prolong its life and avoid limescale deposits.

Obtain disinfectant for swabbing benches and cleaning up spillages
A dedicated marked beaker for a specific disinfectant and a marked waste container make the dilution easier and minimises the risk of over dilution (*e.g.*, to use the disinfectant Hycolin at 2.0% (v/v) make a ring mark at 20 cm^3 on the beaker and a ring mark at 1000 cm^3 on the waste container).

Obtain non-absorbent cotton wool or plastic caps for plugging vessels
Differently coloured cotton wools (or white with a spot of coloured dye) or plastic caps to help to identify different culture media

Obtain aluminium foil or greaseproof paper for protecting cotton wool frombecoming damp during autoclaving and therefore an ineffective barrier to microorganisms
Secure the greaseproof paper over the cotton wool plug with two elastic bands in case one band breaks during autoclaving.

Obtain an inoculating loop for subculturing
New loops can be formed from 32 gauge Nichrome wire using a small pair of pliers.

Obtain Pasteur pipettes for transferring broth cultures
A sufficiently accurate measuring device can be made by connecting a 1 cm^3 syringe to a sterile Pasteur pipette by a short piece of rubber tube; the volume of air drawn up into the syringe is equal to the volume of the liquid drawn into a Pasteur pipette, the contents of which remain sterile because its end is plugged with cotton wool

Obtain McCartney or Universal bottles for keeping slope and broth cultures
Just before performing a transfer procedure check that the cap of the bottle is not so tight that it cannot be unscrewed easily by the little finger

Obtain Petri dishes for preparing agar plates
It is easier to pour molten medium into a Petri dish that is placed near the edge of the bench - and the result is usually a better poured plate

Obtain culture media for growing microbes
A bottle of sterile medium can be conveniently melted in a microwave oven but first check that the non-metal cap of the bottle is slightly loose otherwise the bottle could explode!

Now purchase microbial culture
and light the Bunsen burner

RESOURCE LIST

Microbiology texts and videos
CCAP *(1989) Microbial Engine - Algae and Protozoa* (video). Ambleside, Cumbria: Institute of Freshwater Ecology.

Freeland P. (1991) *Microorganisms in Action*. Sevenoaks: Hodder & Stoughton.

Ingle M.R. (1986) Microbiology and Biotechnology. In: *Basil Blackwell Studies in Advanced Biology*. Oxford: Basil Blackwell.

Lowe P. and Wells S. (1991) *Microorganisms, Biotechnology and Disease*. Cambridge: Cambridge University Press.

Taylor J. (1990) *Microorganisms and Biotechnology* (in: *Bath Science 16 - 19 project*). Walton-on-Thames, Surrey: Nelson and Sons Ltd.

Wymer P. (ed.) (1990) *Biotechnology in Practice*. Cambridge: Hobsons.

Practical Microbiology

Belcher H. and Swale E. (1988) *Culturing Algae*. Ambleside, Cumbria: Institute of Freshwater Ecology (CCAP).

Philip Harris Ltd (1985) *Yeast Growth: Fermentation* (computer software). Shenstone, Staffs: Philip Harris Ltd.

Wymer P. and Grainger J. (1987) *Practical Microbiology and Biotechnology for Schools*. Edinburgh: Macdonald Educational.

Safety guidance

ASE (1988) *Topics in Safety*. Hatfield, Herts: Association for Science Education.

ASE (1992) *Be Safe!: Some Aspects of Safety in Science and Technology in Primary Schools*. Hatfield, Herts: Association for Science Education.

CCAP (1988) *Culture Kit for Practical Microbiology and Biotechnology*. Ambleside, Cumbria: Institute of Freshwater Ecology.

CLEAPSS (1995) *Laboratory Handbook*. London: Brunel University (available only to members of CLEAPSS).

Committee on Safety in Microbiology (1991) *Safety in Microbiology (A code of practice for schools and non-advanced further education)*. Hamilton: Bell College of Technology (copies from Mr. B. Powlesland).

Department of Education and Science (1990) *Microbiology: an HMI Guide for Schools and Non-Advanced Further Education*. London: HMSO.

Verran J. (1991) *An Introduction to Practical Microbiology* (video and teaching pack). Manchester: Manchester Metropolitan University.

Suppliers of Cultures

Philip Harris Ltd
Lynn Lane
Shenstone
Lichfield
Staffordshire WS14 0EE
Telephone 01543 480077

Blades Biological
Cowden
Edenbridge
Kent TN8 7DX
Telephone 01942 850242

Information and Advice

Microbiology in Schools Advisory Committee (MISAC)
c/o Institute of Biology
20-22 Queensberry Place
London SW7 2DZ
Telephone 0171 851 8333

National Centre for Biotechnology Education (NCBE)
Department of Microbiology
University of Reading
Reading RG6 2AJ
Telephone 01734 873743

Society for General Microbiology (SGM)
Marlborough House
Basingstoke Road
Spencer's Wood
Reading RG7 1AE
Telephone 01734 885577

Choose Microbiology
A SGM information leaflet on what microbiologists do and why pupils should consider microbiology as a career.

CCAP Institute of Freshwater Ecology
The Ferry House
Ambleside
Cumbria LA22 0AP
Telephone 015394 42468

CLEAPSS School Science Service
Brunel University
Uxbridge UB8 3PH
Telephone 01895 251496
(For members of CLEAPSS only.)

Scottish Schools Equipment Research Centre
24 Bernard Terrace
Edinburgh EH8 9NX
Telephone 0131 668 4421

Chapter 3 PLANTS

N. Collins and R. Price

Plants support almost all the other life on Earth, directly or indirectly. In many parts of the world plants are the major component of peoples' diet. Planting schemes in inner cities soften stark outlines of buildings and many people have an emotional link with trees in their surroundings. Indeed, the most popular pastime in Britain is gardening, in support of which there is a diversity of magazines and of programmes on television.

The work of plant scientists is of increasing importance. This includes the challenges associated with provision of food, medical care, and energy to sustain an ever-increasing world population without causing lasting damage to the planet. For example, classical plant breeding is now being augmented by methods which overcome genetic barriers and accelerate the production of new crop varieties. As we start to appreciate the importance of maintaining genetic diversity and to see that herbal medicines may have merits, many of the plants from which modern crops are descended, and from which herbal remedies are derived, are threatened with extinction. Fossil fuel reserves are finite; photosynthesis, by contrast, is an ever-present generator of carbon compounds and trapper of solar energy. These and other challenges can only be tackled if we educate, inform and enthuse children about plants. Science teachers all over the world must cultivate a positive and optimistic approach to these plant-centred global problems.

And yet it seems that all too often plants are neglected in school science in comparison with animals (Honey, 1987) and under-represented in science education journals, such as *School Science Review* and the *Journal of Biological Education*. On the whole, plants that have been considered in the biology and science curriculum and suggested as experimental material in textbooks are drawn from a relatively restricted number of species ... geranium, Canadian pond weed, broad bean seeds, tomatoes and mustard and cress. Oats may make a brief appearance, more often than not only as seedlings. Things are changing and many schools and colleges associated with the Science and Plants for Schools (SAPS) program have found that rapid-cycling brassicas are a useful addition to the plants which can be grown successfully in the school laboratory or greenhouse (Price, 1991; Tomkins and Williams, 1990).

Students of Biology on post-16 courses may study a selection of native species during field courses, but since few of the species concerned are likely to be grown within their schools, there is no dynamic interaction with the life cycle of the plants. The same is true of crop plants; few schools grow them from seed to harvest. Even sixth formers can be amazed that carrots and onions flower! Many school sites have ornamental plantings and much mown grass, with the occasional meadow, but since the decline of Rural Science courses, cultivation and plant husbandry within the curriculum has become rare.

A commonly held view is that children in school are not interested in plants - botany is boring. This is paradoxical, because many students are turning to vegetarianism and many supermarkets now carry shelves of exotic fruits and vegetables from all corners of the earth. The western world is supposedly a 'greener' place and yet human biology is where the greatest enthusiasm is likely to be found amongst school children.

Although there are many more university applications to read animal sciences than there are to read plant science, there are probably more career opportunities in plant science than there are in animal science.

Chapter 3   Plants

This chapter aims to demonstrate that it is possible to grow and exploit a variety of plants within schools, indoors and out, which can make a significant contribution to the science curriculum (and beyond). At a time of great curriculum change, which places great demands on teachers, it is important to ensure that any work with plants will give maximum yield, of a curricular nature, for minimum input; much of what we suggest satisfies this requirement.

Experimental work involving animals, beyond basic observations of growth and behaviour, is not an option in many schools. On whatever scale plants are grown, they lend themselves to the sort of investigatory work required increasingly by both pre- and post-16 science courses, whether in the laboratory in studies of seed germination, or in an experimental plot within which, for example, varieties of potato have their performance compared.

After dealing with general matters, including sources of material and facilities for growth and maintenance, cultivation in the laboratory and beyond will be considered through a series of case studies.

SOURCES

While most teachers in England and Wales are aware of the implications of the Animals (Scientific Procedures) Act of 1986 for animals brought into school or used in school science, some may be unaware of the very considerable protection given to plants by the Wildlife and Countryside Act of 1981. It has far reaching implications for teaching of plant biology. Unless you have a licence it is illegal intentionally to pick, uproot or destroy any *specially protected plants*. The collection of seed is also illegal. Most specially protected plants are very rare and unlikely to be encountered. The list, known as Schedule 9 of the Act, is available from English Nature, the Countryside Council for Wales or Scottish Natural Heritage. Teachers in Northern Ireland should refer to the Department of the Environment for Northern Ireland.

For day-to-day botanical activities a more important aspect of the Wildlife and Countryside Act is that it is also illegal intentionally to uproot or pick any plant grown in the wild unless you have permission from the landowner. If you have permission, it is then a matter of applying common sense. Where there are large numbers of a given species, destructive sampling of parts of the plants or even of whole plants is unlikely to threaten the population. For small populations care should be taken that sampling does not threaten their survival.

Some plants are poisonous or can elicit an allergic response (see page 105).

SEEDS

There are a great many more outlets for seeds nowadays than even a few years ago. As well as seed merchants and local garden centres, many large supermarkets and DIY stores sell the seeds of common flowers and vegetables. Health food stores often sell the seeds of many plants for consumption - but many are viable. The addresses of major seed suppliers are given at the end of the chapter - whether or not you buy through this route, their free catalogues are full of fascinating information. Suppliers of scientific equipment and materials to schools have short seed lists and supply some crop seeds in small quantities but if you wish to grow arable crops on an experimental basis it is probably better to contact a local agricultural seed merchant. Seed obtained from seed merchants is often treated with pesticide dressings. If buying them loose by weight, enquire of the seed merchant about this. Read notes on seed packets and take appropriate precautions.

If you need particular types of seed in bulk and their genetic constitution is of no consequence, as for example with broad beans for dissection or germination, then it is worth growing some plants and harvesting your own seed.

WHOLE PLANTS

Many plants are now supplied in containers and are to be found in a great diversity of outlets, including high street chain stores, supermarkets and DIY shops - it can be worth keeping your eye open for plants which have been reduced in price. After a period of convalescence, they often reward their rescuers. Families of many school pupils and, indeed, some pupils themselves will be active gardeners and are often willing to part with propagated or propagatable materials. It is worth contacting them through a blanket letter home or through the school newsletter. There is little reason for ever needing to buy geraniums.

FRUIT AND VEGETABLES

Fruits and vegetables from all over the world are to be found in the greengrocery sections of many local supermarkets - 'cultivating' the manager can lead to cheap or free provision of useful materials, especially once they have reached their sell by date. Another useful supplier can be local allotment holders; it is not uncommon for them to overproduce some crops. Peas are an example and, where this is the case, interesting projects can be undertaken, as described later. It is highly likely that some members of staff will be vegetable growers who may be able to help. Someone within the school is likely also to have green fingers where house plants are concerned.

SCHOOL GROUNDS AS A BOTANIC GARDEN/ARBORETUM

The common weeds which are likely to be found growing all over a school site are an enormously rich potential resource. Many grow very quickly and have interesting survival strategies. Much information is available about the use of school grounds generally, from organisations such as the Council for Environmental Education, English Nature, Learning through Landscapes, and the National Association for Environmental Education (see address list at the end of this chapter).

Planting schemes, especially for trees, can capture the imagination of school children and parents. Whenever planting of new trees is undertaken it is worth considering less common species, such as walnut, yew, *Davidia involucrata* or *Ginkgo biloba*. Trees like rowans and *Buddleia* will attract birds and butterflies. Lime trees are useful for the aphids that sap-suck from their leaves, including those low down on the trees, where children can tag their own leaf and follow its changing fortune, from unfurling in spring to senescence in the autumn. Native species such as oak, ash, birch, beech and Scots pine have a place as well as exotics. Where families wish to mark their child's time at the school in some way, a tree donation is one way to start a mini-arboretum.

School grounds could become important sites for the conservation of rare species. For example, of the world's 662 known conifer species, 364 are listed as threatened by the International Union for Conservation of Nature and Natural Resources (IUCN). Many of these grow in temperate climates and in other parts of the world and can thrive in the British Isles. A good example is the dawn redwood (*Metasequoia glyptostroboides*). This species first became known to western scientists when it was reported in a remote area of China. This was of great interest because the genus *Metasequoia* had been known only from the fossil record in Japan and it was thought that there were no living members of the genus. The tree was first introduced into cultivation in Europe in 1948 and it soon became obvious that it would thrive in many parts of the British Isles, especially in Southern England. This tree is ideal for school grounds. It grows reasonably rapidly but 40-year-old specimens have a spread of only about 2 metres. Unlike many other conifers it is deciduous and therefore has colourful autumn tints.

Pond margins and damp corners of the school grounds can act as sources of algae, liverworts, mosses and ferns. Without necessarily dwelling overlong on the alternation of generations with younger children, there is great beauty and intricacy of structure and function to be revealed with the aid of a dissecting microscope. Movements of moss peristomes and fern sporangia in response to changing humidity are worth investigating. A Biology Culture Kit containing a range of cultures of algae (and protozoa), with a number of suggested investigations, is available from the Culture Collection of Algae and Protozoa, which has produced the kit in conjunction with the Shell Education Service (see Address List).

FACILITIES

Outdoors

Many schools have developed or are actively developing environmental policies and these are likely to encompass cultivated ground and glasshouses, as well as management practices.

It is recommended that any cultivated ground is surrounded by paving slabs. Where plots are dug within a school field, maintenance of neat edges can be made easier by laying as little as a 30 cm wide strip of slabs - mowing machines can pass over them. In primary schools or in situations where only limited numbers and varieties of plants are being grown, a chequerboard, in which alternate slabs or some other pattern are removed from a paved area, can provide a very accessible means of examining plant growth. Within larger plots it is well worthwhile using relatively light 60 cm x 30 cm slabs to subdivide them. These allow all-weather access to plants in the plot and can be re-positioned from season to season, as well as removed temporarily if and when plots are dug right through. It is possible to grow many plants out-of-doors in containers, ranging in size from the largest planter down to the plastic mushroom trays used by supermarket chains, usually available free if you ask.

Glasshouses can be very useful, either in the main growing season for plants such as tomatoes, aubergines, peppers and cucumbers or, with thermostatically controlled heating, for maintaining more delicate plants over winter. Protected cultivation can be expensive; automation of heating, watering and ventilation is essential. Any new glasshouse being erected is best built with aluminium framing, to reduce maintenance costs, with polycarbonate sheet in place of glass, and should incorporate automatic ventilation, which may include fan assistance. Installation of some artificial lighting and mist-propagation units can greatly extend possibilities for project work.

Polythene tunnels can be a lot cheaper than glasshouses, especially if erected by yourself, and can be much larger. Installing a wide central slabbed pathway allows a whole class to be brought inside to work on protected plots. In a polythene tunnel some form of (automatic) watering by mist jet is essential. If facilities are not already automated it may be worthwhile exploring (project-based) assistance from the school's CDT department in the development of sensing and control systems.

All plant cultivation generates waste and this is best composted, perhaps supplemented by vegetable waste from the school kitchen, straw from stables or livestock markets and any plant material picked up by contractors who are grass-cutting on the school site, shredding prunings or collecting dead leaves. Apart from the advantage of recycling materials, much project work can be centred around a well-run compost heap, as is described more fully below. Most school sites generate plant waste in quantities sufficient to keep two or three large compost bins on the go; the product can be dug in or, if sterilised in a soil steriliser, can be used as a growth medium or mulch the following year. It might be worthwhile to invest in a shredding machine.

Each school will have to look to its own security arrangements. Some schools without any fencing beyond that of the school boundary suffer no vandalism; other plantings hidden beyond many lines of defence can fall victim. It may be a help to make the garden a feature of the school into which the local community is welcomed, so that a wider sense of possession is developed, beyond those pupils directly involved.

Indoors

Many plants can be grown in the school laboratory but it is important to understand the limiting factors. The most important are likely to be space, light, water (both provision and loss from transpiration), pollinating agents and temperature extremes (especially in winter). Light must be adequate, both in amount and quality, especially during the winter. Some form of artificial light is therefore essential if plants are to be maintained in a healthy, actively-growing condition throughout the year. Many plants, including rapid-cycling brassicas, will grow successfully under a bank of ordinary cool white fluorescent tubes. However for some plants these tubes may provide insufficient red light and this can be remedied by replacing one or more of the cool white tubes with special 'triphosphor' or 'colour' tubes. Low cost designs for light banks are available from the SAPS project. *Lighting for Horticultural Production* (p44) should be consulted if you are interested in technical information about horticultural lighting.

Watering systems which use wicks and/or capillary matting with large reservoirs help to overcome the problem of long weekends. Plants deprived of sufficient water may not actually wilt but will suffer reduced growth and development. Plants grown on sunny window sills are likely to experience temperature extremes, particularly in winter. High temperatures and high light intensity will accelerate transpiration in many cases. Low temperatures at night, when the central heating is off, will slow down metabolic processes and the plants may begin to show stress symptoms. For an excellent and much more comprehensive account see the *Laboratory Handbook* (CLEAPSS, 1992) pp 1558-1565. Artificial pollination may be necessary if seed is to be collected.

MAINTENANCE

Outdoors

Maintenance of outdoor plots can be very difficult because of the tendency to revert to weeds very quickly, although weeds have many uses when they grow in the right places. Some crops are effective at smothering weeds, once plants are established (*e.g.* potatoes and sugar beet). Mulching, whether with well-matured compost, in which temperatures reached during its production have killed most weed seeds, or with black polythene, can be useful amongst established plants or between rows. Fallow plots can be covered up with weighed-down black polythene until the next stage of cultivation. In the absence of technical or grounds staff support, it is essential to elicit help from keen gardeners, perhaps members of the PTA, or other members of staff. Fifteen minutes weeding at the right time by an interested adult who is a reliable gardener can make a vast difference to the success of projects with plants out of doors.

Even if a facility for growing plants under protected cultivation is fully automated, maintenance can be time consuming and labour intensive but, as mentioned above, possibilities for project work are much enhanced.

Indoors

Some of the plants grown indoors will be the subject of projects run by individual students and responsibility will rest with them, supported by teachers and technicians. There are many places inside schools where plants can be grown. Most schools will have a green-fingered houseplant person on the staff - a lunchtime supervisor, a librarian, a teacher, a caretaker, an ancillary - or an involved parent. Whoever they are, they should be cultivated!

CULTIVATION AND EXPERIMENTAL WORK

Outdoors

There are many advantages in growing plants out of doors. Students can gain direct experience of time scales, seasonality and the entire life cycles of organisms. Unless there is a risk of vandalism, many plants can grown with the minimal attention, presenting no problem during school holidays, whereas those grown under artificial conditions are likely to need constant monitoring and care to ensure their survival, unless systems are fully automated.

The following account includes brief details of a range of projects underway in a small garden at a comprehensive high school in the West Midlands (Figure 3·1). The aim is always to maximise the impact and minimise the maintenance of any project undertaken. A number of the activities are run as major group enterprises, involving students of a wide range of ages and abilities in maintenance and data collection. Interpretation is then carried out at a variety of levels, appropriate to each individual. Our experience with growing many different plants has evolved into a selection which fits the school year well, notably the summer holiday gap. This includes plants which can be harvested before the summer holiday or which can be left to their own devices over the holiday and harvested in the autumn term. Within the garden concerned there has been investment in additional facilities such as an observation window, set into a low retaining wall and behind which seeds can be sown to follow the development of roots below ground.

Any of the plants below may be harvested destructively at some points in their life cycle. Portable digital (kitchen) balances, costing about £40 or so from normal retail outlets, are very useful, since they can be used to determine fresh weights as soon as a plant has been uprooted or some part harvested. Plants may be cut up into component parts (*e.g.* roots, stems, leaves, flowers, fruits or others, such as developing potato tubers or root nodules on legumes), so that partitioning between these components can be examined. Dry weights are obtained by putting plant material, often in bulk, into a drying cabinet with sliding doors, costing between £300 and £400 from scientific suppliers, possibly less when bought as a 'pie-warming oven'. With larger items of fresh plant material some sub-sampling may be necessary. A useful piece of equipment available is an integrating solarimeter, which logs total incoming radiation, so that energy fixation efficiencies of crops can be assessed.

In the following account, based on particular crops, only a selection of the experiments carried out, or possible, is described.

Figure 3·1. A range of projects in a small garden.

KEY

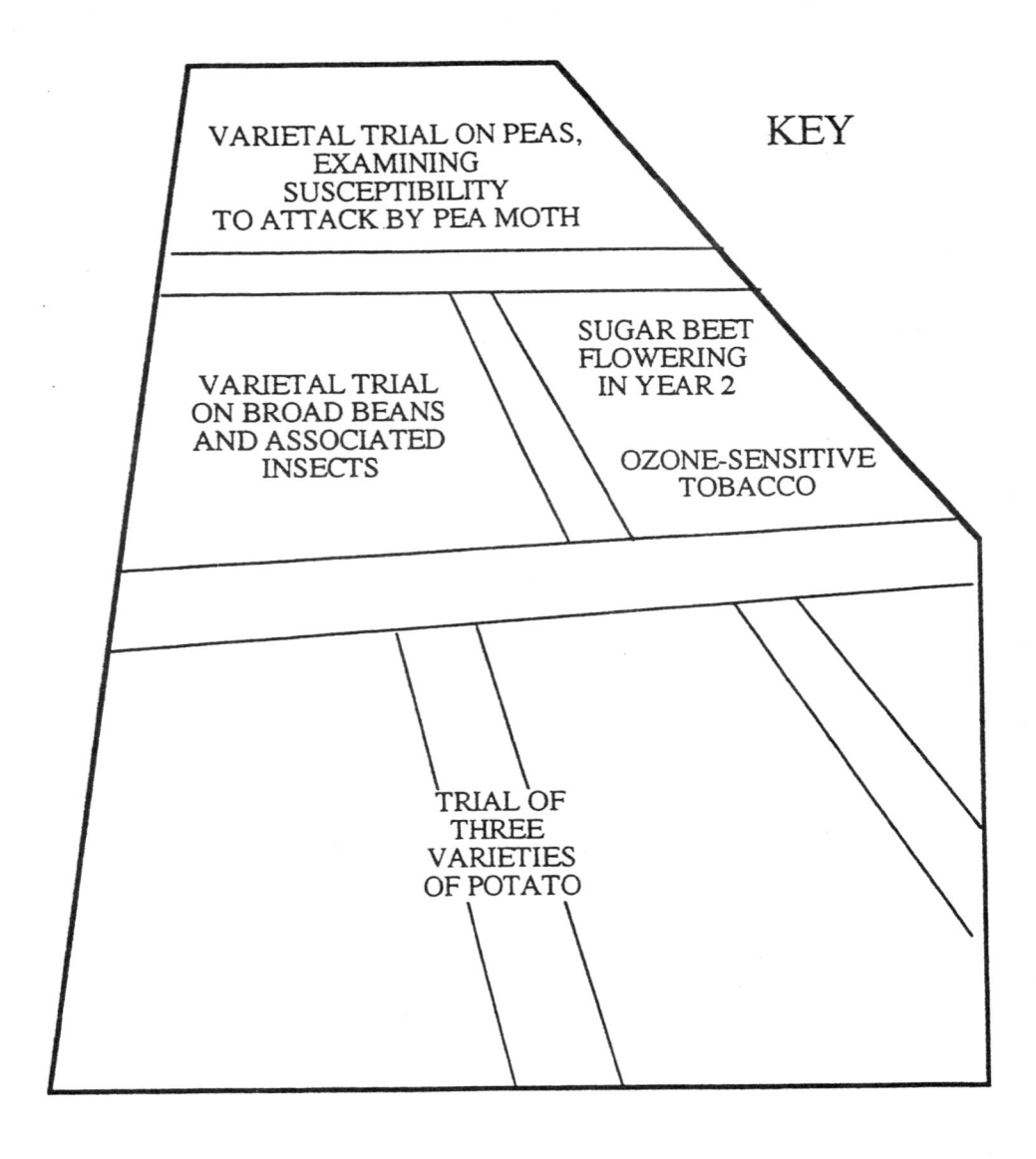

- VARIETAL TRIAL ON PEAS, EXAMINING SUSCEPTIBILITY TO ATTACK BY PEA MOTH
- VARIETAL TRIAL ON BROAD BEANS AND ASSOCIATED INSECTS
- SUGAR BEET FLOWERING IN YEAR 2
- OZONE-SENSITIVE TOBACCO
- TRIAL OF THREE VARIETIES OF POTATO

IN SOIL, UNPROTECTED

1. *Sugar beet - an important root crop which supplies much of Britain's sugar.* Sugar beet is a biennial plant, sown in March/April and harvested between the end of September and December of its first year, for processing in local factories. When fully grown the root may weight up to 5 kg; leaves are large, robust, simple and spinach-shaped. When the seeds are sown directly into the ground in March, they tend to germinate at the same time as many similar looking weeds - a nightmare to contend with if any number are being grown. Farmers can overcome this with suitable herbicide spraying regimes but for the school garden it is easier to sow the seed in trays for subsequent transplanting into weed-free soil. By the time weeds germinate sugar beet are easily identified, have a head start and can soon smother weed growth. Seeds can be obtained, in pelleted form, from British Sugar Plc (see address list at end) and can be sown in soil, in mushroom trays from a local supermarket, topped with about 2 cm of compost to suppress weed growth. Projects undertaken have included:

- Effects of plant spacing. A higher than normal density (10cm x 10cm) produces lots of small roots, which may have the same productivity as those planted at field spacings. Field spacings (20 cm x 20 m) require less seed and there is space between rows for machinery to move through the crop and to maximise harvesting efficiency. Wide spacing (40 cm x 40 cm) yields the largest beet but, with gaps between them, productivity is reduced. A plot of ground 4 m by 2 m can accommodate this activity. A number of different groups, from year 9 through to year 13, are involved at harvest and weighing time, with the data being pooled for analysis.

- Growing plants for the full biennial cycle, with destructive harvesting at intervals. A few plants (or even single plants) are harvested at monthly intervals and cut up into three components, root, petiole and lamina. Each component is weighed and leaf areas measured.

2. *Peas and pea moths - peas are grown to count the caterpillars inside the pods and assess damage.* Peas are sown in short rows, as early as possible in the year, so that the pods are ready to harvest by July. Pods are opened and the peas counted, frequency of pea moth caterpillars determined and the way in which they feed examined. Correlation between degree of infestation and the pod size and factors affecting attack, such as varietal differences, are examined. This can be done with pea pods grown in school or obtained from some other sources.

3. *Broad beans - grown as a varietal trial and to explore insects associated with the crop.* Broad beans are robust plants which can be sown in the spring or in autumn to overwinter. There is a good range of variation between commonly available garden varieties in obvious characteristics, such as height. Data sheets that are a map of the plot containing the plants, with a box for each plant, are easy to use and the plants are robust enough to be visited and measured by many groups of children during a season. Data gathering is by different pairs of students on successive visits, working methodically through the crop, calling out values to the rest of the group, who make their own copy onto the data sheet/ plot map. An alternative is to record directly into a spreadsheet using a palmtop/laptop computer. Data recorded in the school garden, clearly showing the effect of genotype, are given in Figure 3·2.

Because black bean aphids (*Aphis fabae* Scop.) almost always colonise broad bean plants, this system can form the basis of many experiments by students, which can be conducted relatively quickly and with limited apparatus (Backhouse, 1972; Llewellyn, 1984), during the summer term. At a simple level, boxes on the broad bean data sheet can be used also to record insects present - ants, weevils, black bean aphids and their predators, such as ladybirds (adults, mating adults, eggs, larvae, pupae) and hoverfly larvae, are easy to recognise and record - and, again, varietal differences are interesting to investigate. Various parisitoids, including small wasps, can attack the aphids, laying eggs inside them. The larva eats the aphid alive, pupates inside the residual exoskeleton and the adult wasp emerges through a neat 'trap door' which it cuts through the abdominal wall of the blackfly. Careful inspection may reveal hinged-open 'trap doors', from which the extent of infestation can be recorded easily. Useful information about these and other aphid predators is provided by Rotheray (1989). A handy reference text about ladybirds is that of Majerus and Kearns (1989).

Figure 2. Growth of two varieties of broad bean, planted before and after the winter.

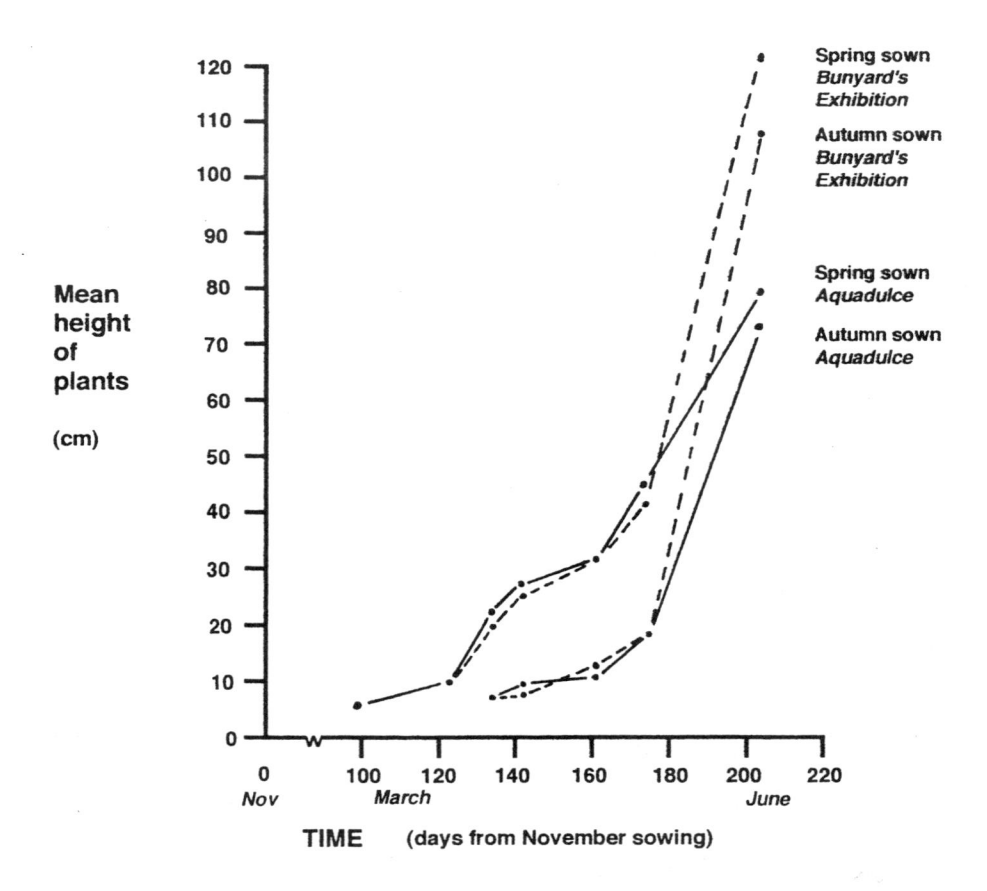

4. *Creeping buttercups - many metres of runners and hundreds of offsets in a season from one plant.* Plants have been grown in soil and in temporary sand culture beds, with added nutrients, where there has been a need to extricate root stems. The sand was put on top of a slabbed area, with polythene sheeting raised up over a low surround of bricks. This is especially suitable for spreading, stoloniferous plants, such as creeping buttercups, and has the advantage that in the first season there is no growth of other plants. During a season many metres of runners are produced, with a distinct pattern of space-filling by hundreds in a season. Shading of runners is likely to lead to less branching and longer internodes - a topic ripe for investigation.

5. *Rhizobium and legumes - plants with root nodules that are easy to grow.* A number of different legumes are always grown, to demonstrate an important part of the nitrogen cycle.

6. *Root growth - seeds sown behind transparent windows in the soil allow tracing of root distribution.* This has been done with sequential sowings of radish, broad bean and, in the case of potato, includes the development of stem tubers. In the garden in question a custom-built glass window has been let into a wall retaining a raised bed. In the absence of such a facility, large plastic lemonade bottles, cut off below the shoulder and with holes in the bottom for drainage, can be filled with soil. Seeds can be planted against the sides of the bottle, the bottles can be sunk in open soil and retrieved when required for viewing roots.

7. *Potatoes - varietal trials examined by a whole year group.* Although one set of children in year 10 have overall responsibility for planting and managing a trial plot of potatoes, harvesting of a number of individual plants is carried out by each year 9 science set, as part of a lesson on growth of food. Two hundred and sixty children participate directly in this longer term project, for about 20 minutes per set, during the autumn term. The data collected are collated and analysed by the prime-movers, by now in year 11. Biologists on post-16 courses may also contribute to data collection and subsequent analysis. (See also the Potato Marketing Board's *Potato Projects - potatoes in the National Curriculum.*)

PROTECTED

1. *Tomatoes - trusses on tomato plants provide distinct locations for data collection and novel ways of presenting growth data.* Tomatoes are easily grown from seed. They can be grown in gro-bags or in soil in the glasshouse. There are very distinct varietal differences in, for example, growth rates, size of fruits, numbers of fruit produced, coloration of fruit. It is possible to produce a vertical map of a tomato plant; each truss is easily identifiable, so that a pupil can relocate it again and again and log the state of development, through flowering to the mature fruit.

2. *Ozone sensitive tobacco.* It is necessary to germinate the seeds under protected conditions - this can be difficult. Once plants are established in pots they can be planted out of doors. Although the original WATCH scheme required the plants to be monitored over a short period they are worth growing on through the summer.

3. *Rice - an example of a tropical crop.* Rice has been grown in 'paddy fields' in old sinks in the greenhouse. It needs to be sown early with heat and extra lighting if it is to be harvested in a temperate summer.

PROCESSING AND DECAY

1. *Silage additives - to improve the pickling process, various preparations are added to grass intended for silage.* Mini-silage clamps can be made out of 40 cm long pieces of 4" (10 cm) plastic soil pipe. Grass which has wilted for a short time after cutting is packed in, sealed and weighed down. It is possible to obtain small sachets of the various additives used commercially to accelerate the process, including enzymes and acid, and to investigate their action.

2. *The magnificent (compost) heap - decomposition and respiration on a grand scale.* Large piles of garden waste material are piled up quickly, perhaps with additional straw and manure (now wash your hands!). Various temperature probes are inserted into and above the heap and an electronic thermometer used to measure temperatures. Automatic data loggers can be useful as well, although this removes from the students much worthwhile responsibility for keeping regular records. Temperatures rise rapidly over a week or so, to as much as 70 °C, a graphic demonstration of microbial respiration. Weed seed in 35 mm film cans can be buried within the heap as it is built, to see whether the temperatures attained are sufficient to kill them.

COMMERCIAL PRACTICES

1. *Propagation from seeds and cuttings - students grow bedding plants, pot herbs and hardwood cuttings for sale within the school community or on contract to a local nursery.* The school, at which most of the projects described here have been undertaken, has strong links with a large wholesale propagating nursery, whose managing director has contracted pupils to supply pot herbs from seed and root cuttings. Eight hundred pot herbs have been supplied to full commercial specification, having been grown within a small glasshouse. The students concerned were working on modules within the City and Guilds Diploma of Vocational Education Foundation unit course, pre-16, although such a link would work well also with post-16 courses, such as GNVQ Science (Advanced).

2. *Mycorrhizae and plant propagation - the local propagating nursery tried out inocula of vesicular arbuscular mycorrhizal fungi to improve striking and growth of cuttings.* Sixth form 'A'-level Biology students conducted a proper large scale trial for the nursery, with cuttings of *Buddleia globosa* grown in commercial modules.

INDOORS

A diversity of work is possible indoors. In this section we consider a range of topics in plant science which are worth tackling on a practical basis with living plants.

THE TROPIC RESPONSE OF SEEDLINGS

It is easy to forget that when a plant responds in some way to light it is often responding to one particular wavelength and the presence of different coloured pigments in plants provides a clue to this. A selection of good filters is a useful resource and if small seedlings are chosen and grown in small containers, such as black 35 mm film cans, the need for large (and expensive) pieces of filter can be avoided. Tropic responses to light are rapid and easily measured and there is plenty of scope for quantitative predictions. The independent variables include temperature, wavelength, light intensity, angle of incidence of light, duration of illumination (and whether continuous or intermittent). The dependent variables could include such quantitative responses as angle, rate and direction of curvature etc. The hypocotyls and radicles of 5-day old seedlings also give geotropic responses. Which has the faster growth rate and which has the more sensitive geotropic response?

In addition to tropisms, several other plant responses to light, especially to the wavelength, provide good topics for higher level investigations, e.g. hypocotyl elongation, anthocyanin synthesis, chloroplast development, stomatal size and density etc. In each case a number of quantitative independent variables can be identified and, with a little ingenuity, ways of making quantitative measurements of the effects of the independent variables can be devised. (Atteridge, 1990; Firn, 1990; Wilkins, 1991). Day length controls on flowering can be investigated also.

DISPERSAL OF SEEDS OR FRUITS BY WIND

The fruits of sycamore, maple, ash, lime and many members of the family *Asteraceae (Compositae)* e.g. dandelion, thistle and sow thistle, are said to be adapted for wind dispersal. Collection of the fruits or seeds, observations of their behaviour and careful examination of their structure will lead to many questions. For example: in winged fruits such as those of sycamore or maple, is the time taken for the fruit to fall a given vertical distance directly proportional to the area of the wings? Why do some winged fruits rotate clockwise as they descend from the tree while others, from the same tree, rotate anti-clockwise? Do winged fruits always fly the same way up?

GENETICS AND ARTIFICIAL SELECTION

The problem is to find a suitable organism with which to work. Rapid-cycling brassicas (fast plants) have several characteristics which make them suitable for investigations in genetics and artificial selection in schools and colleges. The plants are small but robust. With a suitable light source, they can flower 14 days after the seed is sown and complete their life cycle, from seed to seed, in only 5 weeks. There is no dormancy, so another generation can be grown immediately.

Several mutants are available, but little is known of their inheritance patterns. Most of the mutants and the wild type are self-incompatible, so those plants chosen as female parents do not have to be emasculated nor do they need to be anaesthetized while you examine them! The plants are therefore ideal for investigations in genetics (Price and Harding, 1993). They are also very useful for studying the effects of artificial selection. For example the density and distribution of trichomes (hairs) on the leaf surfaces, leaf margins, petioles and stems are very variable. Are these differences genetically determined? Recurrent selection for high and for low trichome density produces significantly different populations after only two generations. Other characteristics, such as anthocyanin expression, are equally interesting. Further information about rapid-cycling brassicas is available from SAPS. Rapid-cycling brassica kits are available from Philip Harris Ltd. Other species, such as the garden pea, are worthy of consideration (Ambrose, Brown and Taylor, 1993).

GERMINATION AND GROWTH

Seed germination is a process amenable to many investigations. Kordan (1992) describes a simple, low-cost experimental system. Why don't tomato seeds germinate inside a tomato? Is there something in the ripe tomato which inhibits germination? If so, does it inhibit the germination of other seeds? Do other fruits have a similar germination inhibitor (Gill, 1982)? The seeds of many plants have two cotyledons.

What happens if you cut off a bit or even a whole cotyledon? Does this affect the growth of the seedling?

It can be interesting to challenge children to come up with a method for the production of mustard and cress on a commercial scale - how many seeds are needed per plastic pot? Is it a commercially viable option to use seed bought through normal retail outlets? What are the ideal conditions for germination? Are these the most economical or profitable conditions?

MINERAL NUTRITION

Many inexpensive and widely available seeds are suitable for studies of plant mineral nutrition. For example, radishes can be grown easily in black film cans under fluorescent tubes. At room temperature they produce a good crop after 2-3 weeks (Hewitson and Price, 1994). What happens if rapid-cycling brassica seedlings are grown without any nutrients, *e.g.* do the plants flower if they have no fertiliser? If they do, can they also set viable seed?

PHOTOSYNTHESIS

Leaf or cotyledon discs will float in 0.2M sodium hydrogencarbonate solution but sink if the air is 'sucked' out of them by reducing pressure (*e.g.* in a plastic syringe). They rise again if exposed to light (and sink if placed in darkness). How can these observations be interpreted? What effect does variation in wavelength or light intensity have? Do discs cut from cotyledons behave differently from discs cut from leaves? Compare a number of fast growing plants *e.g.* brassica, sunflower, radish, white mustard.

What pigments are present in plants? Paper chromatography is a standard technique but thin layer chromatography is worth trying (Tomkins and Miller, 1994). Is it possible to relate the pigments identified to the known action spectrum for photosynthesis?

TRANSPIRATION

Apart from dyes being taken up through the xylem in celery or white carnation flowers, Quicke and Tunstead (1991) have described a method for permanently staining the xylem system within a leaf. Stomata are easy to measure and count using nail varnish impressions. These impressions can be mounted permanently, so that a useful reference collection can be built up. What factors affect the distribution and size of stomata? How does this tie in with rates of water loss from leaves in simple weight potometers?

POLLEN, POLLINATION AND FERTILISATION

How long does pollen stay viable? Under what conditions? Freshly collected pollen from most plants will germinate and grow in a pollen tube if it is placed in a suitable medium. Such media typically contain sucrose and various salts. Does pollen from one type of plant germinate in the same medium as pollen from another plant? How fast do pollen tubes grow? How long must a pollen tube grow in order to achieve fertilisation in the flower? A simple method for studying the growth of pollen tubes is available from SAPS.

Rapid-cycling brassicas are normally pollinated by insects. Which is the best agent for pollination, a 'bee stick' (a dead honey bee glued to a cocktail stick), a paint brush or a cotton wool bud? How would you measure success?

EMBRYOGENESIS AND SEED DEVELOPMENT

In many plants, development of the embryo can be followed from very early stages by dissecting out ovules. What factors influence the development of an embryo? In plants which produce pods, is there an optimum number of pods per plant? How many seeds are there in each pod? What factors influence the number? Are the seeds in a given pod, or on the same plant, all the same size? Are big seeds better than small seeds? What is meant by better? Does position in the pod confer any advantages or disadvantages on the seed, e.g. are seeds at one end bigger or better than seeds at the other end?

PETS AND PATHOGENS

Which parts of a plant do aphids choose to attack? Do caterpillars favour particular leaves? Can ladybird larvae or other carnivorous insects be used for biological control of aphids? Can extracts from "distasteful" plants protect tasty plants? Foe example, brassicas have a characteristic taste. Are brassica extracts toxic to insects or other potential pests? Are they toxic to other plants? Why don't aphids attack tomatoes (Kitching, 1984; Whittaker, 1992)?

PLANT PRODUCTS

Plants produce many products. Many plants are processed in some way before the main product, such as an oil or sugar, is collected. 16% of the fresh weight of a sugar beet root is sucrose - it is a challenge to try to extract this. Other plants produce more exotic products - for example, the meat-tenderising properties of pineapple can be assessed (Dickson and Bickerstaff, 1991).

POLLUTION

Apart from the sensitivity to ozone of some cultivars of tobacco, mentioned in the outdoor section above (p34), controlled laboratory experiments can be set up to investigate the possible effects of acid deposition on plants (Horsley, 1990; Perkins, 1992). Radish and mustard can show effects.

CARNIVOROUS PLANTS

Carnivorous plants can make interesting specimens to show to students (Barker, 1981; Temple, 1988). Many do well under light banks.

A FOOTNOTE

Plant biology is entering a new era, in which there will be rapid development in novel areas. Genetic manipulations for a variety of outcomes are in their infancy. Plants have always provided renewable energy sources but research activities directed at photosynthesis may lead to uncoupling and direct harnessing of the energy-trapping biochemical reactions involved. Whatever the future holds, we hope that through an active engagement with plants in schools, some of your students might be inspired to work with them in their life beyond school.

REFERENCES

Ambrose M.J., Brown C.R. and Taylor P.N. (1993) New pea genes for the classroom. *School Science Review*, **75**, pp 87 - 90.

Attridge T.H. (1990) *Light and Plant Responses*. London: Edward Arnold.

Backhouse M. (1972) A study of blackfly (*Aphis fabae* Scop.) on broad bean plants, and of insects associated directly or indirectly with the colony. *Journal of Biological Education*, **6**, pp 239 - 249.

Barker J. (1981) Carnivorous plants. *Journal of Biological Education*, **15**, pp 15 - 18.

CLEAPSS (1992) *Laboratory Handbook*. Uxbridge, London: Brunel University (available only to members of CLEAPSS). pp 1542 - 1558.

Dickson S.R. and Bickerstaff G.F. (1991) Pineapple, bromelain and protein hydrolysis. *Journal of Biological Education*, **25**, pp 164 - 166.

Farm Energy Centre (1993) *Lighting for Horticultural Production: a guide to the application and design of lighting systems for enhancing horticultural productivity*. Kenilworth: Farm Energy Centre.

Firn R.D. (1990) Phototropism. *Journal of Biological Education*, **24**, pp 153 - 157.

Gill J. (1982) A study of germination inhibition in fruits. *Journal of Biological Education*, **16**, pp 162 - 163.

Hewitson J. and Price R. (1994) Plant mineral nutrition. *School Science Review*, **76**, pp 45 - 55.

Honey J. N. (1987) Where have all the flowers gone? - the place of plants in school science. *Journal of Biological Education*, **21**, pp 185 - 189.

Horsley A. (1990) Acid rain in the laboratory: an investigation into the effects of sulphur dioxide gas on wheat seed germination. *Journal of Biological Education*, **24**, p 71.

Kitching G.B. (1984) Chemical warfare amongst plants. *School Science Review*, **65**, p 499.

Kordan H.A. (1992) Seed viability and germination: a multi-purpose experimental system. *Journal of Biological Education*, **26**, pp 247 - 251.

Llewellyn M. (1984) The biology of aphids. *Journal of Biological Education*, **18**, pp 119 - 131.

Majerus M. and Kearns P. (1989) *Ladybirds*. Slough: Richmond Publishing Co.

Perkins P. (1993) The effect of sulphur dioxide on the germination and growth of cress seeds. *School Science Review*, **72**, pp 90 - 92.

Price R. (1991) Perfect plants for projects. *Biological Sciences Review*, **4**, pp 32 - 36.

Price R. and Harding S. (1993) Genetics in the classroom. *Journal of Biological Education*, **27**, pp 161 - 164. Quicke D.L.J. and Tunstead J.R. (1991) A simple, one-step method for producing permanent preparations of dye-filled xylem systems in leaves. *School Science Review*, **74**, pp 80 - 82.

Roberts E. (1991) How do crops know when to flower? The importance of daylength and temperature. *Biological Sciences Review*, **3**, pp 2 - 7.
Rotheray G.E. (1989) *Aphid Predators*. Slough: Richmond.
Temple P. (1988) *Carnivorous Plants - a Wisley Handbook*. London: Cassell (for the Royal Horticultural Society).
Tomkins S.P. and Miller M.B. (1994) Thin layer chromatography of leaf pigments. *School Science Review*, **75**, pp 69 - 72.
Tomkins S.P. and Williams P.H. (1990) Fast plants for finer science - an introduction to the biology of rapid-cycling *Brassica campestris (rapa) L. Journal of Biological Education*, **24**, pp 239 - 250.
Whittaker J.B. (1992) Green plants and plant-feeding insects. *Journal of Biological Education*, **24**, pp 239 - 250.
Wilkins M. (1991) How does your garden grow? Auxin - what it is and how it works. *Biological Sciences Review*, **3**, pp 22 - 26.

OTHER USEFUL BOOKS

Bell A. D. (1991) *An Illustrated Guide to Plant Morphology*. Oxford: Oxford University Press.
Bingham C.D. (1977) *Plants* (Educational Use of Living Organisms project of the Schools Council). London: Hodder and Stoughton.
Bleasdale J.K.A., Salter P.J. *et al.* (1991) *The Complete Know and Grow Vegetables*. Oxford: Oxford University Press.
De Rougemont G.M. (1989) *A Field Guide to the Crops of Britain and Europe*. London: Collins.
Edlin H.L. (1975) *Collins Guide to Tree Planting and Cultivation*. London: Collins.
Godfrey W. (1986) *Fruit and Nuts - a Sainsbury Guide*. Cambridge: Woodhead-Faulkner.
Hessayon D.G. (1985) *The Vegetable Expert*. Waltham Cross (Herts): pbi Publications.
Hessayon D.G. (1989) *Vegetable Jotter*. Waltham Cross (Herts): pbi Publications.
Howes F.N. (1974) *A Dictionary of Useful and Everyday Plants and their Common Names*. Cambridge: Cambridge University Press.
IOB (1990) *Exploited Plants - Collected Papers from Biologist*. London: Institute of Biology.
Langer R.H.M. and Hill G.D. (1982) *Agriculture Plants*. Cambridge: Cambridge University Press.
Lewington A. (1990) *Plants for People*. London: Natural History Museum.
Lock R. (1993) Use of living organisms. In ASE: *Secondary Science Teachers Handbook* (ed. Hull R.). Hemel Hempstead: Simon Schuster. pp 179 - 205.
Lotschert W. and Bees G. (1983) *Collins Guide to Tropical Plants*. London: Collins.
Mitchell A. (1974) *A Field Guide to the Trees of Great Britain and Northern Europe*. London: Collins.
Salt B. (1990) Growing plants in school. *Journal of Biological Education*, **24**, pp 103 - 107.

ADDRESS LIST

Suppliers of seed
General catalogues:

    Suttons Seeds Ltd
    Hele Road
    Torquay
    Devon. TQ2 7QJ
    Telephone: 01803 612011

    Samuel Dobie and Son Ltd
    Broomhill Way
    Torquay
    Devon. TQ2 7QW
    Telephone: 01803 616281

    Thompson and Morgan Ltd
    London Road
    Ipswich
    Suffolk. IP2 0BA
    Telephone: 01473 688821

Agricultural Crops

Further information on the crops and limited quantities of free seed can be obtained from these addresses:

Sugar beet:

    John Price
    British Sugar Plc
    PO Box 26
    Oundle Road
    Peterborough
    PE2 9QU

Cereals:

    Homegrown Cereal Authority
    Hamlyn House
    Highgate Hill
    London
    N19 5PR

    Flour Advisory Bureau Ltd
    21 Arlington Street
    London
    SW1A 1RN

    Kellogg Company of Great Britain Ltd
    Stretford
    Manchester
    M32 8RA

Potatoes:

    Information on potatoes can be obtained from:
    Potato Marketing Board
    Broadfield House
    4 Between Towns Road
    Cowley
    Oxford
    OX4 3NA

'Seed' potatoes can be obtained locally, but if you are after particular varieties obtain a catalogue and then order from:

>Edwin Tucker and Sons Ltd
>Commercial Road
>Crediton
>Devon
>EX17 1ER

Algae (and protozoa)
A culture kit for practical microbiology and biotechnology, with either culture of algae and protozoa, culture media, booklets on identification and culture, with a guide for teachers and technicians. The kit can be obtained from:

>The Administrative Officer, CCAP
>Institute of Freshwater Ecology
>The Windermere Laboratory
>Ambleside
>Cumbria
>LA22 0LP
>Telephone: 015394 42468

Rapid-cycling brassicas
Full details, including how to construct a light bank can be obtained from:

>Science and Plants for Schools
>Homerton College
>Hills Road
>Cambridge
>CB2 2PH
>Tel: 01223 411141 Ext. 233

Horticultural Crops
Details leaflets about many horticultural crops, as well as pests and diseases affecting them, are available from:

>The Liaison Officer
>Horticultural Research International
>Wellesbourne
>Warwick
>CV35 9EF
>Tel: 01789 470382

48 Chapter 3 Plants

Carnivorous plants
A useful source of plants and information, from whom a catalogue and price list can be obtained is:

>   Marston Exotics
>   Brampton Lane
>   Madley
>   Hereford
>   HR2 9LX
>   Tel: 01981 251140

ADVICE ON DEVELOPING SCHOOL GROUNDS:

>   Council for Environmental Education
>   School of Education
>   University of Reading
>   London Road
>   Reading
>   RG1 5AQ
>   Tel: 01734 756061
>
>   English Nature
>   Northminster House
>   Peterborough
>   PE1 1UA
>   Tel: 01733 340345
>
>   Learning through Landscapes Trust
>   South Side Offices
>   The Law Courts
>   Winchester
>   Hampshire
>   SO23 9DL
>   Tel: 01962 846258
>
>   National Association for Environmental Education
>   University of Wolverhampton
>   Walsall Campus
>   Gorway
>   Walsall
>   WS1 3BD
>   Tel: 01922 31200

Chapter 4 ANIMALS

M Cassidy and J. Tranter.

The benefits of keeping animals in the classroom have long been recognised. An opportunity to examine live animals close at hand is important both for effective teaching and for pupil motivation. It can make real an otherwise lifeless textbook account and also engenders an appreciation and concern for all forms of life. The Science National Curriculum states that, at Key Stage 1, children "should have opportunities whenever possible through first hand observation to find out about a variety of plant and animal life"; similarly pupils at Key Stage 2 "should investigate some aspects of feeding, support, movement and behaviour in relation to themselves and other animals". Studies of animal life extend into secondary schooling with pupils having to "broaden their study of locally occurring plants and animals" and relate behaviour to survival and reproduction. There is a need therefore for schools to attend to the problems of maintaining small (often temporary) animal cultures for use in school. Invertebrate animals (*i.e.* those not possessing a backbone) are particularly suitable in this respect.

INVERTEBRATES

Although the educational benefits of studying vertebrates and invertebrates are similar, there are important differences. Vertebrates are larger, more familiar and more "accessible" to pupils, especially younger ones. Invertebrate animals demonstrate a much wider diversity of types, a greater range of form and function and are relatively easy to collect and maintain. They are less demanding in their requirements than vertebrates and generally pose fewer health risks. Invertebrates can be highly attractive organisms and will readily stimulate pupil interest, indeed many of the specimens brought into school can be from pupils themselves.

The collection and transfer of animals from the wild into the classroom is accompanied by particular responsibilities. Animals must be handled sensitively, and over collection or collection in environmentally sensitive areas avoided. There are legal restrictions regarding what may and may not be taken (DES Memorandum 3/90, Animals and Plants in Schools: Legal Aspects, see page 85) but in general most native invertebrates may be collected. The teacher responsible needs to take account of housing and feeding (remember weekends and holidays) and must be conscious of pupil safety as well as animal welfare. Although zoonoses (diseases of animals communicable to humans) are rare in invertebrates there is still a small risk of bites, stings, allergies, cuts and water-borne diseases. Pupils must be encouraged to handle invertebrates sensibly, to clean out tanks regularly and to wash their hands thoroughly. It is sensible to begin with simple, relatively undemanding organisms such as stick insects, woodlice, slugs and snails and to expand as experience is gained. A mixed collection is preferable (native and exotic/terrestrial and aquatic) which can be housed adequately in the classroom/laboratory if attention is paid to staging (cages and tanks on level benches with easy access), sinks (close proximity), windows (avoid draughts, avoid direct sunlight) and storage (to house the animals, store food, collect soiled materials, etc.).

The collection and maintenance of invertebrates does not require a lot of equipment (e.g. glass and plastic containers, white tray, a variety of nets). Two types of habitat will probably be most fruitful for collecting: a body of fresh water (pond, lake or stream) and a mixed woodland (invertebrates in the leaf canopy, leaf litter, in and around fallen trees, etc.). However, do not forget the garden (Chinnery, 1978) which can still provide a most important reservoir for wild-life. In a more urban setting, flower beds, old walls and unmown grass can provide a variety of insect life.

Invertebrates can be used in a wide variety of classroom investigations in such areas as

- life history studies
- biometrics (growth studies, egg laying)
- behavioural studies
- studies of diversity and adaptation
- principles of classification
- genetics
- ecological (habitat) studies
- population studies
- studies of anatomy (internal and external appearance)
- studies of physiology

Such investigations may be used for the assessment of a number of skills such as observation, measurement, manipulation and deduction both as short-term and long-term (project) investigations. There is still much to find out regarding the habits even of some of our commonest invertebrates; schools and colleges are well suited to this task.

SIMPLE, SOFT-BODIED INVERTEBRATES

Hydra belongs to the Phylum Cnidaria, which also includes the jellyfish, corals and sea anemones. It is a common freshwater animal and can be located in almost any permanent body of clean water. *Hydra* are small (15-30 mm in length), simple, soft-bodied animals, with slender bodies surrounded by 4-8 contractile tentacles arranged in a single ring. They are found attached to water plants, submerged wood and stones, and use stinging cells on their tentacles to catch and subdue prey. They are of interest because of their extremely simple body structure, unusual method of prey capture and an ability to reproduce both sexually and asexually.

Green and brown *Hydra* must be kept either in the water from which they were collected or in a specially prepared culture medium - tap water is generally unsuitable. Tap water can be used, however, if left for several weeks with algae or other aquatic plants growing in it. Its suitability can then be tested by placing a few specimens of *Hydra* in the water and observing their reactions. If they extend fully the water is suitable; in unsuitable water the column remains contracted and the tentacles fail to expand. The presence of plants is not necessary in *Hydra* culture but the water does need to be kept cool (no higher than 20 °C) and the culture should be covered.

Hydra will feed on *Daphnia* (slow moving with weak resistance), brine shrimp larvae or tiny particles of lean meat. When placed in a watch glass or other shallow dish, previously starved animals (for 24 hours) can be observed feeding. It is preferable to use a binocular microscope or hand lens for this activity. For a closer examination of stinging cells a dilute solution of sodium chloride will stimulate discharge of these cells in the *Hydra* tentacles. Asexual reproduction occurs by budding, the young *Hydra* "buds" breaking away from the parent when fully formed. This can build up local populations with surprising rapidity. Sexual reproduction occurs seasonally with ovaries and testes seen externally on the body wall. Sex organs when wanted for class display may be induced in most species of *Hydra* (notably brown *Hydra*) by placing the culture at a temperature of between 10 °C and 15 °C for two to three weeks. The culture should be fed regularly.

Flatworms, sometimes called planariums, are small (usually less than 10 mm) free-living soft-bodied animals belonging to the Phylum Platyhelminthes. They are flattened dorso-ventrally (top to bottom) and move by secreting mucus and gliding over surfaces. When disturbed they will show wriggling/swimming movements but out of water they contract to become immobile blobs of jelly. Flatworms are common freshwater animals found among plant debris, beneath submerged stones and on the underside of floating leaves (*e.g.* water lilies) often in large numbers. *Crenobia alpina* occurs on stones in rushing upland streams while *Polycelis nigra* (a dark-coloured planarian with a row of eyes arranged around the anterior margin) is common in lowland lakes and streams. Most are carnivorous; the cream-coloured *Dendrocoelum lacteum* (common in many lowland lakes and ponds) feeds largely on the freshwater louse while *Dugesia polychroa* (with a rounded head and pronounced "neck") feeds upon water snails.

Flatworms can be collected from suitable ponds and streams either by baiting with liver or raw beef (submerged in a glass jar) or by bringing in quantities of submerged vegetation and sorting. The animals can be rinsed from their substrate using pond (not tap) water and kept in cool conditions. They may be handled using a small clean paint brush. Flatworms can be maintained successfully in the laboratory if fed regularly with some sort of fresh meat in clean pond water. They are particularly susceptible to fouling and the water must be changed regularly. Restlessness and excessive mucus secretion may suggest too much chlorine in the (tap) water. If the stock is merely being maintained, it is sufficient to feed once or twice a week, removing any uneaten food after three to four hours.

Suitable investigations might include, for instance, in flatworms previously starved for 48 hours observations of klinokinetic behaviour: moving towards the food and testing the strength of the chemical stimulation on either side by means of lateral movements of the head. A piece of graph paper stuck to the underside of a Petri dish will allow the direction and rate of movement to be determined accurately. By placing a mirror beneath the Petri dish it is possible to see the animal feeding (protrusion of the pharynx). The path the animal takes can be recorded using a felt-tipped pen on the Petri dish lid; alternatively, the mucus track made by the animal during its wanderings may be highlighted by sprinkling with talcum powder on the base after the contents of the dish have been removed. Response to light (unilateral illumination or the effects of light intensity on rate of locomotion) may be tested in the Petri dish bearing in mind the inverse square law (half the distance away from the light source, four times its light intensity) and the probable increase in temperature! Planarians may also be used to investigate growth under different feeding regimes and responses to heat and cold. In keeping with many invertebrates, flatworms can regenerate body parts following damage. It is worthwhile to keep a look out for such animals.

Earthworms belong to the Phylum Annelida - the segmented worms. They are cylindrical animals with around 150 segments bearing small bristles of chaetae on the under surface with which to grip the soil. They are generally reddish brown in colour and sexually mature individuals possess a reproductive structure, a prominent band, about a third of the way down from the head (the more pointed end!). There are about 25 native species with two of the commonest being *Lumbricus terrestris* and *Allolobophora longa*. *Lumbricus* is more red brown in colour and does not produce the typical worm casts (soil is voided in the upper reaches of their burrows). *Allolobophora* is abundant in most gardens between October and May (it goes into diapause during the summer), has more of a dirty brown coloration and makes typical soil casts.

Earthworms can be collected locally or may be purchased from a reputable supplier. They can be found in most rich garden-type soils, particularly under old grassland and orchards where the soil has been left undisturbed but they avoid acidic, waterlogged and sandy soils. They can be dug up with a garden fork or spade (avoid damage) or may be extracted by applying dilute solutions of potassium permanganate to the soil. Solutions of detergent and mustard are also very effective as they also irritate the sensitive skin of the animals, causing them to surface. (Animals extracted by chemical means must be washed immediately in clean water.) Within the classroom/laboratory they may be housed in a purpose-built wormery (Rothamsted pattern, Nuffield Biology Teachers Guide I) but this is now discouraged as they are not especially tolerant of the warm dry conditions indoors. It is sensible to keep earthworms outdoors in a large, sturdy container, in cool, moist (30% soil water) conditions. They can be fed on most organic material, leaf litter, etc. including dung (apparently horse dung is best!) which is spread over the surface. A fairly light soil should be used, including a fine sprinkling of sand, and the culture examined every couple of weeks for cocoons (this also helps to aerate the soil). Earthworms do not climb well but a lid is a sensible precaution.

Earthworms are excellent animals to use for observing burrowing. The action of the chaetae and the powerful muscular waves of contraction can easily be seen with a hand lens. Their sensitivity to vibration and light allows the animals to forage at night, or in wet weather, but retreat to their burrows at the onset of day (avoiding predators and the drying sun). Earthworms can be used in investigations involving food preferences (*Lumbricus* pulls leaves into its burrow), respiration (carbon dioxide output, heat generation), reproduction (seminal vesicle smear) and excretory structures (microscopic examination of nephridia). They exhibit habituation (learning not to respond to stimuli, e.g. vibration) and also show chemical communication following exudation of coelomic fluid. The effectiveness of a hydrostatic skeleton can be demonstrated through an examination of earthworm locomotion on a horizontal surface. Bristles (chaetae) on the underside can be both seen and felt when the animal is placed on the palm of your hand.

SLUGS AND SNAILS

Slugs, snails, squid, octopus and sea-shore limpets are all animals belonging to the Phylum Mollusca. There is no standard molluscan body plan but one normally thinks of a hard shell, soft body and slippery skin. Molluscs are a diverse group of animals, generally slow-moving and, as a consequence, highly adapted to their particular habitat. Terrestrial forms are particularly suited to classroom use; they can be kept indoors relatively easily and lend themselves to a number of observational studies.

The land snails and slugs are characterised by the possession of a "lung" chamber and complex reproductive behaviour. There are no larval stages in these terrestrial forms and miniature adults emerge after hatching from the egg (complete with shell in the case of snails). Slugs and snails are restricted in their habits by the need to avoid dehydration. They are active mainly at night and in wet weather, an adaption that prevents desiccation and avoids many day-time predators. Most species live permanently near the soil surface, some climb vegetation, others rest in cracks and crevices between rocks. Slugs, although lacking a waterproof shell, are nonetheless very manoeuvrable and can burrow quite deeply into the soil and find the more inaccessible crevices. Many species also hibernate over winter.

Molluscs feed by means of a flexible, rasping radula, a sort of file in the mouth. Most types feed on vegetation, some achieving pest status. Many will eat carrion but only a few are actively carnivorous, *e.g.* the shelled slugs (Family Testacellidae). The diet of land snails and slugs is generally varied, the grey field slug, for example, has been found to contain buttercup, stinging nettle, ivy, grass, moss, earthworm and insect particles in its gut. When in captivity, food can be placed in a shallow dish and removed daily to avoid fouling. Snails require lime or chalk for their shells and this can be given in powder form with their food (not the chalk substitute found in blackboard chalk!) or a lump of natural chalk or limestone will do.

Slugs and snails can be maintained in a cool place indoors, in a glass or plastic tank (old aquarium tank or plant propagator) with a tightly-fitting lid (avoid overcrowding). The tank or terrarium should be well aerated and kept moist (though not *too* damp). Gravel and well-drained soil can cover the floor with leaf litter, stones and pieces of wood completing the semi-natural environment. No species has a restricted diet and nearly all can be raised on lettuce, carrot, dandelion, hogweed, porridge oats, etc. It is a good idea to clean out the tank completely every month and provide new soil and new leaf litter.

Snails and slugs can be collected locally. The common garden snail, *Helix aspersa*, and the white tipped snail, *Cepaea hortensis*, are good species to begin with together with the large black slug, *Arion ater*, and the grey field slug, *Agriolimax reticulatus*.

Sample investigations might include observing locomotion (measuring speed, looking at mucus production, ability to move over smooth and rough surfaces).

Feeding studies can be undertaken. Do snails and slugs have food preferences? The action of the radula (they will feed on the emulsion layer of <u>well-washed</u> photographic negatives, showing radula marks) can be observed. Time taken for food to pass through the gut (examination of faecal pellets containing identifiable food, *e.g.* carrots) may also be investigated.

Simple sensory experiments can also be carried out - responses of the body to touch, light and chemical stimulation (sugar and vinegar "tastes", crushed lettuce, mint and toadstool odours).

A new and exotic mollusc supplied by education suppliers and local pet shops is the giant African land snail, *Achatina fulica*. This animal is kept under similar conditions to our own snail (though it needs to be kept warm). There have been some comments regarding the possible health risks (CLEAPSS, 1992). As with many molluscs there is a small risk from *Salmonella* bacteria but the risk of meningitis-like symptoms is non-existent from stocks bred in the United Kingdom.

INSECTS

All of us are familiar with insects; indeed they make up at least half of all the different species of organisms existing today. They are found in all habitats (except the open seas), exhibit a remarkable diversity of form and function and can be represented by immense populations. The locust or cockroach is often thought of as the "typical" laboratory insect but increasingly stick insects, fruit flies, flour beetles, ladybirds, ants and butterflies are found in schools and colleges. There is no limit to the range of insect types that may be brought into the classroom to illustrate topics as diverse as behaviour, reproduction, feeding, life cycles, ecology, genetics and locomotion. They can be large and easily handled (such as the larger stick insects and caterpillars), small and social nesting (ants and bees) or freshwater aquatic forms. There is a large (specialist and amateur) literature with many organisations willing to provide assistance.

Insects can be collected almost anywhere. A good place to start is a flowering hedgerow on a warm sunny day, but equally profitable might be a grassy meadow or a patch of nettles, each of which will harbour characteristic species. A garden, of course, is often where most young collectors begin. However, the owner might not take kindly to sweep netting or trampling and an area of scrubland can be just as profitable. Woodland collecting can be disappointing unless you know what to expect and where to look for it. In the depths of a wood, the light is poor and there are few flowers; coniferous woodland is especially gloomy. Freshwater habitats, both still and moving waters, can provide a number of different zones where adult and juvenile insects abound. Insects may be caught by <u>netting</u> (sweep netting grassland, catching insects in flight or pond netting), <u>beating</u> (*e.g.* a tree branch and catching the dislodged fauna on a sheet below), <u>trapping</u> (light trapping, pitfall traps) or simply picking up individual insects using forceps, a specimen tube or pooter (Figure 1).

Figure 4·1. Insect trapping.

Insect stocks purchased from educational suppliers are often best kept in specialist cages. A typical locust cage, for example, consists of an aluminium frame and sides with perforated zinc floor and removable glass front panel. Sand-filled tubes are provided for egg laying together with 60 Watt electric light bulbs for heat and light. The cage has small hatches to facilitate feeding and removal of dead grass; it is easily disassembled for cleaning. Animals feed on fresh grass (cut daily) or wheat bran (occasionally). A suitable daytime temperature is 34 °C with the interior of the cage kept relatively dry. The Nuffield biology project has been instrumental in introducing this animal into classrooms as a type insect. Activities might include an examination of mating (including courtship), egg laying (*e.g.* substrate preferences), growth and development (*e.g.* effects of temperature), colour variation (in species with solitary and gregarious forms, *e.g. Locusta migratoria*, feeding methods can include a microscopic examination of locust mouth parts). Some people show allergy problems following exposure to high density populations of locusts. Both staff and pupils should avoid prolonged exposure; this is best achieved by keeping locusts when needed and *not* attempting to breed them all year round. When cages are cleaned out, avoid raising dust; (the use of a vacuum cleaner is often a good idea).

Cockroaches (*e.g. Periplaneta americana*), are relatively easy to rear provided a temperature of 25 °C and a relatively humid atmosphere are maintained (for breeding, a slightly higher temperature of 27-28 °C is preferable). A small glass or plastic aquarium is suitable with tightly-fitting lid, ventilation holes and a thin line of Vaseline placed around the top of the container to act as a barrier to movement.

Sawdust provides an adequate flooring with balls of newspaper and corrugated cardboard at one end for shelter, and food and water dishes at the other end. Cockroaches can be fed on bran, porridge oats, bread and small pieces of lettuce. They do not need to be cleaned out very regularly (providing the food has not gone mouldy) but do require large amounts of water (interesting responses to water are seen). Cockroaches, however, are difficult to handle (they scuttle away rapidly, and are difficult to grip) and are potential pest species establishing themselves rapidly within buildings. They engage in a number of interesting activities but the need to keep cultures of this insect in school must be questioned. Tropical species of cockroach, requiring extra heating *e.g.* giant hissing cockroach, are unlikely to infest if they escape and are slower moving animals.

Caterpillars are the feeding and growth stages in the life cycles of butterflies and moths. They are familiar to most pupils and can be collected either as eggs or larvae. It is important to note the food plant on which they were discovered. Eggs and small caterpillars are best kept in small plastic boxes, they do not need any special ventilation and larger caterpillars can be moved up into larger boxes as they increase in size (generally speaking the eggs darken before hatching). The correct food plant - *e.g.* nettles (small tortoiseshell butterfly), cabbage (large white butterfly), poplar and willow leaves (poplar hawk moth) - must be provided otherwise the caterpillars will wander around aimlessly and eventually die. Stale food and frass (faecal droppings) should be removed regularly, fresh food provided, and excess moisture avoided. As with many insects, caterpillars will drown in open water. Food plants should be in containers with the necks covered or the plant material can be pushed into watered 'oasis' - the material used by florists for flower arrangements. Readiness for pupation is seen in the restless wanderings of the larger caterpillar. Different species have specific requirements and it is advisable to transfer the animals to simple breeding cages that can accommodate the cocoons which may be suspended from plant stems, (small tortoiseshell), pushed up against the wooden sides of fences, and other objects (large white) or found overwintering in the soil (poplar hawk moth - overwinters as a pupa without forming a cocoon). The Lepidoptera (butterflies and moths) are particularly useful for studying life histories, egg-laying, growth and development (including complete metamorphosis), feeding preferences and the efficiency of conversion of leaf material into caterpillar tissue.

Beetles (Coleoptera) constitute the largest of the insect orders with over 3500 British species. They show great diversity in form and function (from tiny black flea beetles to the large pond diving beetle) and many are of economic importance. The flour beetle, *Tribolium confusum*, may be bred in the laboratory to demonstrate principles of genetics (although it has a longer life cycle than *Drosophila*) but by and large the Coleoptera do not appear in many elementary textbooks. They are often difficult to rear and difficult to observe but simple glass or plastic tanks can be used in which these animals can complete their life cycles. Specific requirements include a soil base, rotting tree stumps or sprigs of foodplant, depending upon the species under study.

Earwigs are a fascinating and ancient insect order (the Dermaptera) with five British species. The common *Forficula auricularia* is omnivorous, feeding on both plant and animal remains. The sexes are easily identifiable (the pincers of the male are more strongly curved than those of the female) and are harmless insects in spite of their appearance! Earwigs are active mainly at night and can be easily trapped by providing daytime resting sites, *e.g.* inverted flower pots stuffed with straw or crumpled paper. They also have a fondness for ripening apples. Square sandwich boxes provide an adequate rearing container with a layer of soil 2-3 cm deep in the base. Shelter can be provided by a piece of bark or broken plant pot, food (slices of carrot, apple, lettuce or cabbage, spiders and slugs) may be placed on a watch glass. The container must be kept at a high humidity (wet cotton wool) but not too wet. Food must be removed regularly. Earwigs require little ventilation.

Investigations with earwigs might include their responses to light (note: their directional response does not exhibit the side-to-side head movements typical of the blowfly larvae). They can be tested in choice chambers for a comparison of light and dark, or damp and dry environments. Horizontal and vertical hollow stems might be tested to see which are more efficient at trapping. Unusually for solitary insects, earwigs make good parents. The female is an excellent mother, licking the eggs (keeps them free of mould?), collecting scattered eggs and feeding the newly- hatched young. This aspect of the animals' reproductive behaviour makes a fascinating subject for study.

Other laboratory insect work involves the behaviour of blowfly larvae (maggots) including responses to light, to temperature, to food and contact with solid objects. Stick insects are popular subjects for the study of crypsis (camouflage), locomotion, incomplete life cycles and growth rates (including effects of the environment). They are especially easy to rear and make good classroom subjects for any age group. Fruit flies (*Drosophila* spp.) are important in the study of genetics while comparative studies of insect mouth parts and legs (adaptation to the environment) are commonly encountered in many school text books. Further insect investigations might include aphids (greenfly), ladybirds, leaf-mining and gall-forming insects. The insect microfauna of leaf litter may be examined as might the range of life forms and adaptations seen in aquatic insects (*e.g.* comparison of fast flowing and slow moving streams; use of the surface film).

OTHER ARTHROPODS

The Phylum Arthropoda is an amazingly successful and diverse group containing not only the insects but also arachnids (spiders and mites), crustaceans (shrimps and crabs) and myriapods (centipedes and millipedes). All arthropods share a common design and a long evolutionary history being the only invertebrates to have successfully adapted to a full terrestrial existence and to have become modified for flight. All members of the group show a segmented body pattern, a hard exoskeleton, jointed limbs and growth in stages following a period of moulting. Typically, the woodlouse is the characteristic non-insect arthropod used in schools but increasingly spiders, centipedes and *Daphnia* are used in a range of observational and experimental studies.

There are 42 British species of woodlouse. The pill woodlouse is commonly found in woods and characteristically rolls itself up into a ball when disturbed. The two commonest species though are *Oniscus asellus* and *Porcellio scaber*, both dull grey in colour; the latter has an upper body surface covered by small tubercles. *Porcellio* is particularly common in bark crevices and on old walls and can tolerate drier habitats than *Oniscus*. The breeding season for many woodlice lasts most of the summer. Eggs are laid into a brood pouch with pale tiny woodlice emerging after four to five weeks. They overwinter sheltered in leaf litter or in deep crevices. They can be kept in an old aquarium tank, its base covered with soil (10 cm deep), dead leaves, flat stones and rotting wood. A lid is required to prevent escape and the tank needs to be sited in the shade, out of direct sunlight. The contents should be moist, not wet. Specimens must be handled gently (forceps will often damage the animals) and woodlice can be fed on a variety of food material (decaying animal and plant remains together with young seedlings).

Although they are generally nocturnal, a number of interesting investigations may be carried out using woodlice - patterns of movement, speed of movement on different substrates, food preferences (using leaf discs), responses to light, humidity and contact with solid objects (choice chambers may be purchased from educational suppliers or improvised using Petri dishes).

Centipedes and millipedes, although belonging to different arthropod classes, nonetheless possess similar limitations to those of woodlice - *i.e.* they are confined to moist, dark areas (due to a lack of an efficient waterproof covering) with many living in soil and leaf litter. Centipedes possess one pair of legs per segment and are carnivorous; millipedes have two pairs of legs per segment and generally feed on decaying plant material. Both possess defensive secretions produced by their outer cuticle and show responses to light, humidity and touch similar to woodlice. As with woodlice they may be housed in a simple terrarium. This group has not been particularly well studied and many original observations might be made by school and college groups. Tropical giant millipedes are now quite widely available from pet stores and specialist suppliers.

Spiders are not always the most popular of animals but they are relatively easy to keep, need little maintenance, show interesting reproductive and food capture behaviour and demonstrate an alternative arthropod anatomy. Arachnophobia is, to a large part, learned and the early introduction of these interesting creatures into our schools may possibly allay some of the fears.

Spiders are very numerous (up to 2 million in an English summer meadow) and diverse (nearly 600 British species) occupying a wide range of habits and habitats. They can be kept in any convenient lidded container (Figure 4·2). Two abundant species are the common garden spider (*Aranea diadema*) and the common house spider (*Tegeneria domestica*).

58 Chapter 4 Animals

The former requires damp conditions and supports from which to build its orb web; the latter day, shaded and sheltered conditions. Generally they do not need to be fed more than once a week, (nearly all spiders can be fed on live flies); if the abdomen looks shrivelled the animal is either thirsty or hungry; if it is refusing food it may be about to moult. Classroom investigations include studies of the life cycle and the dispersal of spiderlings, web construction and the microscopic examination of spider silk, courtship, mating and responses to vibrations on the web (including tuning forks of different pitch).

Some larger, exotic spiders are sold in pet stores. These may inflict a painful bite (but rarely do!) and are not difficult to keep. However, unless specialist expertise is available, they are best avoided

Figure 2. Spider investigations.

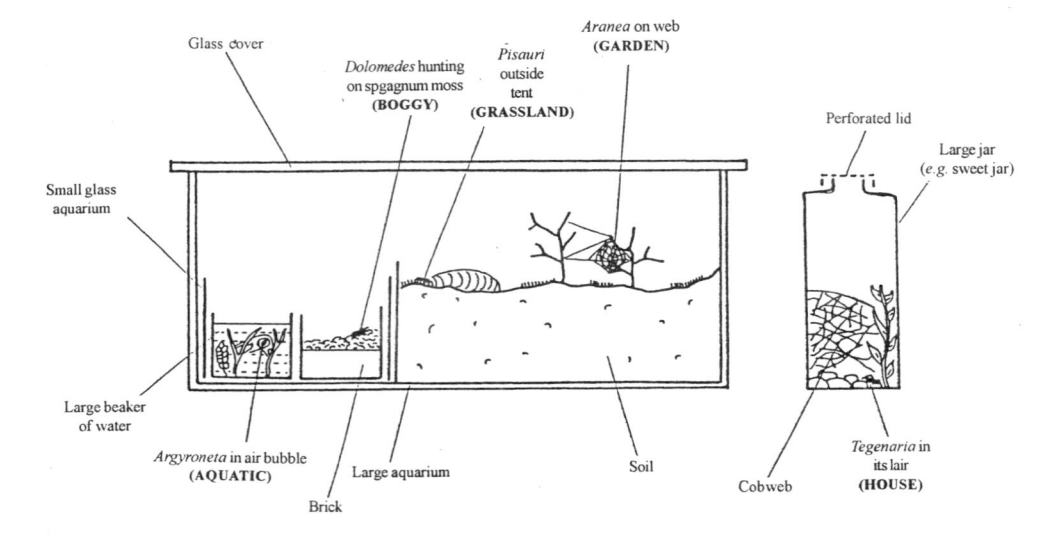

Daphnia species (belonging to the class Crustacea) are increasingly common animals in laboratories. They are useful as food cultures for *Hydra* and aquarium fish but more importantly they are very useful for teaching purposes. Because of their transparency they can be used to illustrate (under the microscope) many anatomical structures (they can be held on a microslide by moist fibres of cotton wool or Vaseline) or to display physiological activities such as feeding, egestion, circulation and reproduction. They also have the advantage of genetic uniformity when reproducing asexually.

It is not always easy to maintain a culture of *Daphnia* in a school laboratory; the type and position of container seem to have an effect on life span. Porcelain containers situated in the shade seem to give best results. There are many weird and wonderful recipes for cultivating *Daphnia* but the essential food for the majority of this group is bacteria or single-celled (not filamentous) algae. They can be maintained at a cool room temperature in containers filled with filtered pond water. Culture (feeding) methods include inoculating the tank with water from an aquarium tank becoming green with phytoplankton; adding a suspension of live yeast until the mixture is slightly milky or Banta's method of a mixture of garden soil, pond water and horse manure! Possible investigations include the effects of chemicals (*e.g.* aspirin) and temperature on heart beat rate, environmental factors (especially oxygen concentration) on their vertical distribution in a water column, response to light and the haemoglobin content of animals kept in poorly and well-aerated water.

It must be remembered that *Daphnia* can easily be purchased in most pet shops (check to ensure they are not mixed with *Cyclops* spp.) and it is often easier to purchase them as and when required.

CONCLUDING REMARKS

The aim of this section has been to introduce the reader to the diversity of invertebrate types and the possibilities afforded by these animals in classroom use. They may be housed simply and cheaply and, preferably, returned to the wild after use. Investigations are many and varied, limited only by the imagination of the teacher, but comparative studies (*e.g.* feeding and locomotion) are particularly instructive. They can be introduced to any age and ability group. A large literature is available (some of which is listed here) but the most important attribute required by the classroom teacher is a sense of curiosity, wonderment and a willingness to experiment.

FURTHER READING

A bibliography is provided for individuals requiring additional information on the culturing of specific invertebrate groups. Journals such as the *School Science Review* and *Journal of Biological Education* are excellent sources of articles on keeping animals in schools. Similarly, CLEAPSS School Science Service produces a number of very useful guidelines on animals in schools (but these are only available to members).

Brock P.D. (1985) The Phasmid Rearers Handbook. *The Amateur Entomologist,* **20**.
Brown V.K. (1983) *Grasshoppers. Naturalists' Handbook 2.* Cambridge: Cambridge University Press.
Bullock C.J. and Street M.L. (1978) Keeping ants in the Laboratory. *School Science Review,* **59**.
Byron M.S. (1988) *How to Keep Stick Insects.* London: Fitzgerald Publishing.
Chinner M. (1978) *The Natural History of the Garden.* London: Fontana.
CLEAPSS (1992) *Giant African Land Snails. L197.* London: Brunel University.
Comber L.C. and Hogg M.E. (1979) *Animals in Schools. Volume 2: Terrestrial Invertebrates.* London: Heinemann Educational Books.
Cooper B.A. (ed.) (1969) Hymenopterist's Handbook (Facsimile edition). *The Amateur Entomologist,* **7**.
Forsythe T.G. (1987) *Common Ground Beetles. Naturalists' Handbooks 8.* Richmond, Surrey: Richmond Publishing Co Ltd.
Hogarth P.J. (1983) Crabs in Labs: the shore crab (*Carcinus meanus*) as teaching material. *Journal of Biological Education,* **17**(2), pp 105 - 111.
Kinchin I.M. (1988) Nematodes in school biology. *School Science Review,* **69**, pp 735 - 737.
Kinchin I.M. (1992) Vacuum sampling of bark dwelling arthropods. *School Science Review,* **73**, p 70.
Majerus M.E.N., Kearns P.W.E., Ireland H. and Forge H. (1989) Ladybirds as teaching aids 1. Collecting and culturing. *Journal of Biological Education,* **23**(2), pp 85 - 95.

Majerus M.E.N., Kearns P.W.E., Forge H. and Burch L. (1989) Ladybirds as teaching aids 2. Potential for practical and project work. *Journal of Biological Education*, **23**(3), pp 187 - 193.

Moriarty F. (1969) The laboratory breeding and embryonic development of *Chorthippus brunneus* (Tunberg). *Proceedings of the Royal Entomological Society of London* (ser. A), **44**(1-3), pp 25 - 34.

Murphy F. (1980) *Keeping Spiders, Insects and Other Land Invertebrates in Captivity*. Edinburgh: John Bartholomew and Son Ltd.

Needham J.G. (1937) *Culture Methods for Invertebrate Animals*. New York: Dover.

Orlands B.F. (1977) *Animal Care from Protozoa to Small Mammals*. London: Addison-Wesley.

Pallant D. (1967) Slugs in school biology. *School Science Review*, **48**.

RSPCA (1985) *Animals in Schools*. Horsham, West Sussex: RSPCA Education Dept.

Walsh G.B. and Dibb J.R. (1954) A Coleopterist's Handbook. *The Amateur Entomologist*, **11**.

Ward-Booth K. and Reiss M. (1988) *Artemia salina*: an easily cultured invertebrate ideally suited for invertebrate studies. *Journal of Biological Education*, **22**(4), pp 247 - 251.

Whittaker C.D. (1988) Keeping *Daphnia* in the laboratory over a long time period. *School Science Review*, **70**, pp 66.

Wilcock H. (1977) Snails in the classroom. *Natural Science in Schools*, Summer 1977.

Williams R.J.A. (1969) A woodlouse vivarium. *School Science Review*, **50**.

Wooton A. (1972) Rearing British Lepidoptera. *Natural Science in Schools*, Spring 1972.

VERTEBRATES

Keeping and studying vertebrates in schools can be most rewarding. Pupils are likely to interact strongly with vertebrates and their presence in the classroom or laboratory will provide opportunities for motivating, investigatory work and observations which can extend those carried out with a variety of invertebrates. Work with selected vertebrates will provide excellent material for studies spelt out in the National Curriculum for Science.

Opportunities to keep and use vertebrates present themselves throughout years 7-13, but it is perhaps in primary school that work with vertebrates is most important. If pupils are to gain a complete and balanced view of the animal kingdom, it is vital that they have the chance to gain first-hand experience of the characteristics and behaviour of vertebrate groups and, as a result, begin to develop appropriate attitudes towards them. Such contact with vertebrates should occur from the earliest years of a pupil's education.

There are many examples of eminently suitable activities with vertebrates which can be tackled well in primary schools. For example, children can be asked to think about pet animals and find out what kinds of vertebrates are kept in people's homes. This will introduce them to the name of a range of vertebrate animals and can be the start of a wider fact-finding exercise when they perhaps visit local pet shops. Pupils will soon see the necessity of putting into groups the various types of vertebrate pets they encounter. Such elementary work on classification can lead on to direct observations of selected vertebrates that are kept in the classroom or brought in temporarily. Comparisons between, for example, the classroom guinea pig and a cat or dog that is brought in to school for a lesson will be most fruitful. Longer-term activities with certain animals can be arranged in the classroom and these could include measuring the growth of a small mammal in relation to the food eaten, incubating the eggs of chickens or ducks and studying the development of common frogs or toads.

Keeping or helping to look after vertebrates, particularly handling small mammals, can also have a therapeutic and emotionally satisfying effect on all pupils. Much work has demonstrated that such involvement is especially beneficial for those who have difficulty in establishing relationships with other pupils and adults or who have special needs. Since many pupils will inevitably have, or obtain in the future, vertebrate pets at home, it is vital that they are helped at school to become responsible, caring and sensitive pet owners. They need to become aware of possible problems that their pets might create, develop an awareness of, and consideration for, their needs and learn the necessary skills in looking after them. All of these will arise naturally out of practical studies of vertebrates at school.

If pupils are to develop the right attitudes to vertebrates, it is obviously crucial that in schools all animals are looked after well and are never neglected or taken for granted. Although the routine care and maintenance of the animals will inevitably involve expenditure of money and time, such costs must be anticipated and always paid in full. Holiday periods pose particular problems but at these times the animals should be cared for as well as usual, whatever alternative arrangements may have to be made.

FISH

In some ways, the most easily kept vertebrates are fish, particularly if only cold-water species such as goldfish are selected. These animals do not require heating and, with natural lighting, equipment requirements are minimal, though filtration of the water is recommended, if not always essential. Tropical species provide a much wider and more interesting variety of fish, in terms of their colour, body form, feeding and behaviour, but they require fluorescent lighting, aeration/filtration and heating of their environment. Tropical marine fish are even more spectacular but too difficult and expensive to keep successfully in most schools.

Apart from showing their general characteristics (as is the case for each group of vertebrates), fish provide many opportunities for investigatory or observational work. A well-balanced tank of tropical community fish can reveal how organisms with apparently similar requirements can live together without intense competition for resources. Different species will be seen moving in different areas of the aquarium, feeding in different ways and on different foods. One does not have to keep 3-spined sticklebacks to study aspects of courtship and other behaviour (though they are excellent for such studies). Much can be achieved, even with goldfish, including investigations of conditioning. Can, for example, a fish be 'trained' to respond to certain external stimuli such as a light being switched on or to feed only from a particular colour of dish? Quantitative measurements of breathing rate can easily be made by counting the pumping movement of water over the gills; this can be related to animal size, activity and water temperature (if both cold and tropical species are kept). Such studies can be extended to investigate oxygen consumption, related to body size or activity, by, for example, analysing water samples for oxygen content using an oxygen electrode or by chemical test such as the Winkler method. Studies of fishes swimming can also lead on to work with models in tanks of water to investigate the influence of fins on locomotory thrust, stability and balance.

There is a wealth of material to which teachers can turn for information when keeping fish. CLEAPSS (1990, 1991) has provided a simple introduction for schools in establishing cold water and tropical aquaria. Mills (1984) is typical of many authors in giving a simple but good account of keeping tropical fish species. More detailed sources include Axelrod and Vorderwinkler (1986).

AMPHIBIANS

A cold-water aquarium can also be used to house a variety of amphibians which are readily available and do well in schools. These include adults which are entirely aquatic: African clawed toads, *Xenopus* spp. and the axolotl, *Ambystoma mexicanum*. *Xenopus* toads come to the water surface to breathe, so aeration of the water is unnecessary. The axolotl, however, breathes through feathery gills, even though adult, so aeration/filtration is desirable. Both animals are carnivorous but it is usually not difficult to provide them with readily available food, for example, strips of liver or heart from the butcher, maggot or mealworms from the pet shop and, with *Xenopus*, even an artificial diet. Leadley Brown (1970) describes in detail the maintenance of *Xenopus*, while Mattison (1992) discusses axolotls, along with *Xenopus* and many other amphibians.

The cold-water aquarium, with water that is constantly filtered, is also the best means of keeping frog and toad spawn (of common, unprotected species) and the tadpoles that subsequently emerge. Only a very small volume of spawn should be taken and kept in a large tank. Too many tadpoles in a small bowl of unfiltered water will inevitably lead to an unnecessarily high mortality. The aim should always to be rear as many tadpoles as possible to adulthood and then return them to the wild. The tadpoles of the American bullfrog, *Rana catesbiana*, are sometimes reared; they are attractive because of their large size and slow growth. However, before deciding to keep such tadpoles, think long term. When they metamorphose into adults, the frogs must be given a permanent home. They must **not** be released into the wild; this is illegal as the bullfrog is not native to Britain.

It is not at all easy to maintain adult British frogs in schools for any length of time. If adult frogs and toads, other than completely aquatic species, are to be kept, then choose North American species such as *Rana catesbiana* mentioned above or the leopard frog, *Rana pipiens*.

Amphibians present an excellent opportunity for pupils to study, easily, the life cycle of a vertebrate. If housed properly, the spawn of common frogs or toads can readily be reared to produce adults. A full cycle can be observed if *Xenopus* adults are injected with gonadotrophin to induce mating, egg laying and fertilisation. (This is an acceptable, legal practice provided the intention is to induce breeding and so continue the colony, but skill is needed in performing the injections.) *Xenopus* is also excellent for studying factors involved in the detection of food material; items of food can be placed in the water at one end of a tank and observations made of the behaviour of a *Xenopus* at the other end. The axolotl is an extraordinary amphibian to keep as it illustrates the principle of neoteny in which the juvenile features are retained in the adult stage; the mature animal is effectively a breeding tadpole.

REPTILES

Reptiles are somewhat more difficult than amphibians to rear in schools but they are very rewarding animals. The myth of reptiles being 'cold and slimy' is easily dispelled by a suitable species of snake in residence. All reptiles require a secure, heated vivarium which is thermostatically controlled, but the need for conditions that are more difficult to provide can be avoided by choosing appropriate species. When purchasing reptiles, they should always be obtained from <u>captive-bred</u> stocks in the United Kingdom.

Many species of lizard need very specialist care but the leopard gecko, *Eublepharus macularius*, is relatively undemanding and does well in schools, even being persuaded to breed if a male is kept with several females. It does not, for example, require U-V radiation which most other lizards must receive to manufacture vitamin D. The leopard gecko feeds readily on easily available live foods such as crickets and waxmoth larvae, though food should be dusted with a vitamin/mineral supplement. It cannot climb the glass sides of a vivarium, so is less likely to escape than other species!

Possibly the least demanding snakes are the garter snakes, including the most common species, *Thamnophis sirtalis*, and the chequered garter snake, *Thamnophis marcianus*. These can usually be fed on earthworms or with whitebait or lancefish (from the local aquarist) which have been boiled for a few minutes to break down an enzyme that destroys essential vitamin B. Garter snakes are, however, adept at escaping, move fast and may suffer from disease more than other snakes. Other recommended species include the corn and rat snakes, *Elaphe guttata* and *E. obsoleta* and the common king snake, *Lampropeltis getulus*. These snakes, however, require rodents as food, though these can be bought frozen, in bulk. All snakes require hides and basking areas and adults need a cool period to simulate winter hibernation.

Reptiles <u>may</u> be carrying *Salmonella* (but so may a chicken from the butcher), so high standards of hygiene at all times are essential. Terrapins, by their aquatic nature, may present more of a risk of infection if care is not taken. They are also <u>not</u> easy to keep for 'beginners' and it must be appreciated that a small specimen can become very large when fully grown. If terrapins are to be kept, it is important that they are adequately housed in a <u>large</u> aquarium in which the water is thermostatically heated. Terrapins require both a water and land area within the tank with a sloping 'beach' between the two; the water need be no deeper than 20 cm. Full spectrum lighting which provides U-V radiation is essential for vitamin D production and to encourage normal behaviour. A basking area beneath a heat source is also required.

All reptiles are useful in teaching about the 'cold-blooded' poikilothermic habit. Observations of their behaviour will reveal that a more or less constant body temperature can be achieved by moving in and out of hotter regions of the vivarium. Studies of the movements of snakes are also always fascinating, particularly how the pattern of locomotion can be affected by the nature of the surface over which the animal is moving.

Mattison (1992) is a good general reference for keeping reptiles, while Mattison (1989, 1991), and Sweeney (1992), provide more details for the care of snakes and lizards. Highfield (1992) provides authoritative advice on rearing terrapins.

BIRDS

Work with birds in schools is most likely to involve the incubation and hatching of eggs of chicken, ducks and possibly other species such as quail. Specialist incubators are required for this task but many schools already possess such equipment. Alternatively, a suitable model may be borrowed from such teachers' centres or even from certain suppliers of incubators. CLEAPSS (1988) and Centre for Life Studies (1989) discuss the use of various models of incubator, the incubation procedure and rearing of hatched birds. Obviously, before such activities are embarked on, it is important for most schools to have identified someone who will take the adults as soon as they have matured sufficiently to be moved outside. Only a few establishments will have the facilities to keep the chickens, ducks, etc. themselves.

Other adult birds, including budgerigars and a variety of small seed-eating species such as waxbills and zebra finches, are best kept in aviaries which will provide the animals with adequate flight room. Trollope (1992) is typical of several authors in giving useful advice on establishing a successful collection of birds, suitable for schools.

The procedure of incubating and hatching eggs successfully is a lesson in itself on the conditions required for normal development within the egg. Careful control of temperature and humidity is needed (the chick must lose some water but not too much), together with regulation of the supply of oxygen and the removal of carbon dioxide. Once the chicks have hatched, there is a host of further activities that are also possible. For example, there is much scope for investigating the chicks' eyesight; they often will peck only at bright objects and ignore food scattered on the floor of the brooder. Observations of the chicks pecking at objects can lead on to studies of their feeding behaviour; as the birds grow older, do they show particular preferences for types of food, perhaps related to its size or colour?

Studies of growth and development are also easily accomplished. The chicks can be weighed regularly and these measurements related to the mass of food and water that the birds consume. Part of the food intake is used in maintaining the chicks' body temperature and this can lead on to studies of cooling and insulation, using model chicks made out of tin cans or bottles, covered in various ways and filled with hot water.

As the chicks mature, much can be discovered in studies of their behaviour and how certain patterns only develop in older birds; suggestions for activities are given in the Centre for Life Studies booklet (1989). If birds other than hens or ducks are kept, such behavioural studies can be extended to include these animals and comparisons made between species.

MAMMALS

Perhaps the most important vertebrates to be kept in schools are small mammals, with which pupils can develop the greatest empathy and from which begin to appreciate the demands and needs of pets at home, including larger mammals such as cats and dogs. The small mammals most likely to be encountered are the mouse, *Mus musculus*, the rat, *Rattus norvegicus*, the Syrian (golden) hamster, *Mesocricetus auratus*, the Mongolian gerbil, *Meriones unguiculatus*, the guinea pig, *Cavia porcellus*, and the rabbit, *Oryctolagus cuniculus*. Less common species include the Russian hamster, *Phodopus sungorus*, Libyan (pallid) jirds and, occasionally (because the animals are very expensive), the chinchilla, *Chinchilla laniger*.

The choice of which small mammals to keep should be determined by considering a number of issues. For example,

* will pupils handle the animals? (Rats, guinea pigs and rabbits are best.)
* is an active animal with interesting behaviour important? (Gerbils score highly; nocturnal hamsters are less suitable.)
* are resources, both in terms of time and money, for routine maintenance limited? (Again, gerbils probably win.)
* will genetic breeding investigations be conducted? (Mice, with some strains of gerbils, have advantages here.)
* is space or size of caging limited? (Rabbits and guinea pigs need a lot of room and, ideally, access to outside pens.)

All small mammals should be obtained from a reputable source to ensure that the animals are themselves healthy and unlikely to transmit disease to humans. The Laboratory Animals Breeders Association (contact c/o the Institute of Biology) operates a register of breeders supplying high quality stock. Schools should note, however, that animals from such sources will be more expensive and that it is unnecessary to invest in the 'ultraclean', 'barrier-bred' small mammals that are available.

There is a wealth of reference material available on the care, housing and handling of small mammals. CLEAPSS (1994) provides clear guidance on how to keep small mammals in schools; the RSPCA (1989) presents alternative views. Wray (1974a,b) gives more detail on individual small mammals and their accommodation needs. For authoritative but very technical information, refer to Poole (1987). (This source also discusses all other groups of vertebrates, as well as small mammals.)

Activities with small mammals can be quite wide ranging. Obviously it is much better to investigate the characteristics of mammals using a live mammal as the teaching aid than it is to resort to theoretical discussions. While observing and handling a small mammal, it will be hard for pupils not to notice the animal's fur, movements of the rib cage, the warm body or the heart beating, along with other features. Such observations can easily be extended to include quantitative studies of breathing and heart rate, relating these to basic physiology, size and surface area. The ideas for work on body temperature maintenance related to heat loss and insulation, discussed in the section on birds, can readily be adopted here. Observations of feeding can be related to the animal's dentition and such work extended to include comparisons with larger herbivores, carnivores and relevant skull material that is available.

Work on human reproduction can often arise naturally from studies of a small mammal breeding colony but the small mammals provide ample opportunity in their own right for work on parental care, the development of newly-born animals and even the oestrous cycle. If suitable animals are chosen, aspects of the inheritance of coat colour are easily tackled. There is also much scope for interesting work on many aspects of behaviour. The booklet on small mammals (Centre for Life Studies, 1984) has two sets of accompanying notes suggesting suitable activities with small mammals in primary schools and secondary establishments.

REFERENCES

Axelrod H. and Vorderwinkler W. (1986) *Encyclopaedia of Tropical Fishes*. Waterlooville, Portsmouth: Nylabone Ltd, TFH Publications.

Centre for Life Studies (1989) *Incubating and Hatching Eggs*. London: Brunel University (copyright acquired by CLEAPSS).

CLEAPSS (1988) *Egg Incubation L71*. London: Brunel University (available only to members of CLEAPSS).

CLEAPSS (1990) *Cold Water Aquaria L181*. London: Brunel University (available only to members of CLEAPSS).

CLEAPSS (1991) *Laboratory Handbook* (Chapter 14.3: Aquaria). London: Brunel University (available only to members of CLEAPSS).

CLEAPSS (1994) *Small Mammals*. London: Brunel University (available only to members of CLEAPSS).

Highfield A. (1992) *The Tortoise Trust Guide to Tortoises and Turtles*. London: Carapace Press.

Leadley Brown A. (1970) *The African Clawed Toad: a Guide to the Biology, Care and Breeding of Xenopus leavis*. Sevenoaks, Kent: Butterworth.

Mattison C. (1989) *Keeping and Breeding Snakes*. London: Blandford Press.

Mattison C. (1991) *Keeping and Breeding Lizards*. London: Blandford Press.

Mattison C. (1992) *The Care of Reptiles and Amphibians in Captivity*. London: Blandford Press.

Mills D. (1984) *Fishkeepers' Guide to the Tropical Aquarium*. London: Salamander Books.

Poole T. (ed.) (1987) *The UFAW Handbook on the Care and Management of Laboratory Animals*. Harlow, Essex: Longman Scientific and Technical.

RSPCA (1989) *Small Mammals in Schools*. Horsham, Sussex: RSPCA.

Sweeney R. (1992) *Garter Snakes: their Natural History and Care in Captivity*. London: Blandford Press.

Trollope J. (1992) *Seed-Eating Birds: their Care and Breeding*. London: Blandford Press.

Wray J. (1974a) *Animal Accommodation for Schools*. Sevenoaks: Hodder and Stoughton (EUP) (for the Schools Council).

Wray, J. (1974b) *Small Mammals*. Sevenoaks: Hodder and Stoughton (EUP) (for the Schools Council).

Chapter 5 PUPILS AS A RESOURCE

P. Horton

ABSTRACT

The human body can provide samples of human tissue which are acceptable alternatives to the use of animal tissues. Physiological exercises can be performed using the human body to monitor breathing rates, heart and pulse rates, and clinical temperature. The range of information thus obtained can be extended using a universal interface, a microcomputer and suitable software. Health studies make a stimulating and relevant way of enhancing the study of biology. Studies involving dietary requirements can relate to digestion, dental decay, and control of body weight. Drugs and their abuse can be linked to their effects on the respiratory, nervous and vascular systems, and discussions on tobacco and alcohol-related diseases. A check list is provided on the ethical guidelines when working with human participants. It is advised that teachers make themselves familiar with these requirements when planning to work with their pupils.

INTRODUCTION

Most teachers support the use of practical work not only as a means of developing pupils' practical skills but also to improve attitudes, foster interest and enjoyment and encourage initiative and self-awareness. These attributes are never better encouraged than when using pupils themselves as a resource.

The use of animals in teaching biology has been increasingly criticised over the last decade not only by those with an animal-rights philosophy but also amongst students of biology (Langley, 1991). There would appear to be widespread concern about the use or killing of animals for educational purposes. One way of reducing or replacing the use of animals in experimental work is to introduce humans as an alternative species. In addition, there are a variety of human-orientated teaching aids which are currently being developed for the use of teachers in the classroom.

This chapter will look at alternatives to the use of animals in school science classes by suggesting that humans are used as a source of tissues and to provide ideas for pupil-based experiments. It will also review currently available aids to teaching physiology and behaviour, other than by direct experimentation. An evaluation of equipment and materials for human physiology investigations is given in CLEAPSS (1993).

HUMAN TISSUES

The body provides valuable material for microscopic studies on cells and tissues. Human hairs plucked from the skin provide a very simple alternative to the collection of epithelial cells from inside the cheek (Tomlins, 1988). Box 5·1 describes how cheek cells may be collected and the Institute of Biology is still of the opinion that this can be done safely providing the procedures are followed carefully. In the past, the DES has advised that it cannot support the methods advocated by the IOB because while accepting that these procedures do not present a risk of HIV infection if followed in some classrooms, it may be that circumstances are such that some pupils could disregard the procedures

However, it could now be argued that if an assessment of risks indicates that the procedure can be followed satisfactorily, then there is no reason why it should not be used. Schools should note, however, that some LEAs do not permit the sampling of cheek cells by any method, and that rules must be obeyed (though they can be challenged and approval sought for the use of a safe method).

Box 5·1. Procedure for the preparation of temporary mounts of human cheek epithelial cells

1. Take a cotton bud from a newly opened pack.
2. Move the cotton bud over the inside of the cheek on one side of the mouth and along the outer lower side of the gum.
3. Smear the cotton bud over a small area of a clean microscope slide.
4. Place the used cotton bud immediately in a small volume of absolute alcohol in a suitable container, (e.g. 5 cm^3 of absolute alcohol in a 10 cm^3 specimen tube).
5. Place 3 drops of 1% methylene blue from a dropper pipette onto the smear and cover with a cover slip.
6. Observe the smear under the low power magnification of a microscope. When the cells are in focus increase the power of the objective to achieve maximum magnification and resolution.
7. After the cells have been observed, immerse the slide and cover slip in a beaker of laboratory disinfectant.
8. The teacher or laboratory technician should place the used cotton buds in a polythene bag which should be sealed and then disposed of in accordance with local regulations governing the disposal of laboratory waste.
9. Slides and cover slips should be washed thoroughly, dried and re-used according to normal practice.

N.B. CLEAPSS prefer 1% minimum strength hypochlorite solution to absolute alcohol.

The plucked hair method is suitable for Key Stage 3 pupils and can be used to observe the outer hair root sheath cells (Wells, 1991). Another suggestion is to apply transparent adhesive tape to the wrist, remove and observe when applied to a microscope slide. Any of these studies would form a good introduction to cell biology, epithelial tissue, the skin, and the relationship of structure to function in animal cells for level 5 of the National Curriculum.

In addition, when stained with an aqueous solution of methylene blue, chromosomes dividing by mitosis may be seen and would stimulate discussion among Key Stage 4 pupils on the part played in the body by cell division and requirements for the continuous renewal of epithelial layers.

Teaching about the composition of blood is not made easy, since both the IOB recommendations (IOB, 1987) and the DES guidance (DES/WOED, 1987) have advised that the procedure for taking blood smears must be discontinued in schools and colleges. Most employers (LEAs) accepted this guidance and have prohibited blood sampling in their establishments. It should be noted, however, that in schools and colleges which are not subject to LEA rules (because they are independent, grant-maintained or incorporated establishments), it may be permissible for blood samples to be taken after an assessment of the risks involved has been made and it is judged that all necessary precautions can be taken to ensure a safe procedure. The structure of blood cells can be taught to 14-16 year olds using models. For example, models of red and white blood cells can be cut out of plastic foam and painted over with gloss paint (Mansell, 1989a). Such an approach can be extended to advanced level studies to explain the ABO blood group system by attaching to the red cells cut-outs of A- and B-shaped antigens and complimentary Y-shaped A and B antibodies (Mansell, 1989b).

PHYSIOLOGICAL EXERCISES

In the past it has been customary to use animals to investigate such processes as breathing and the circulation of blood, however there are plenty of possibilities at all levels for self-experimentation (Figure 5·1).

Figure 5·1. Physiological exercises.

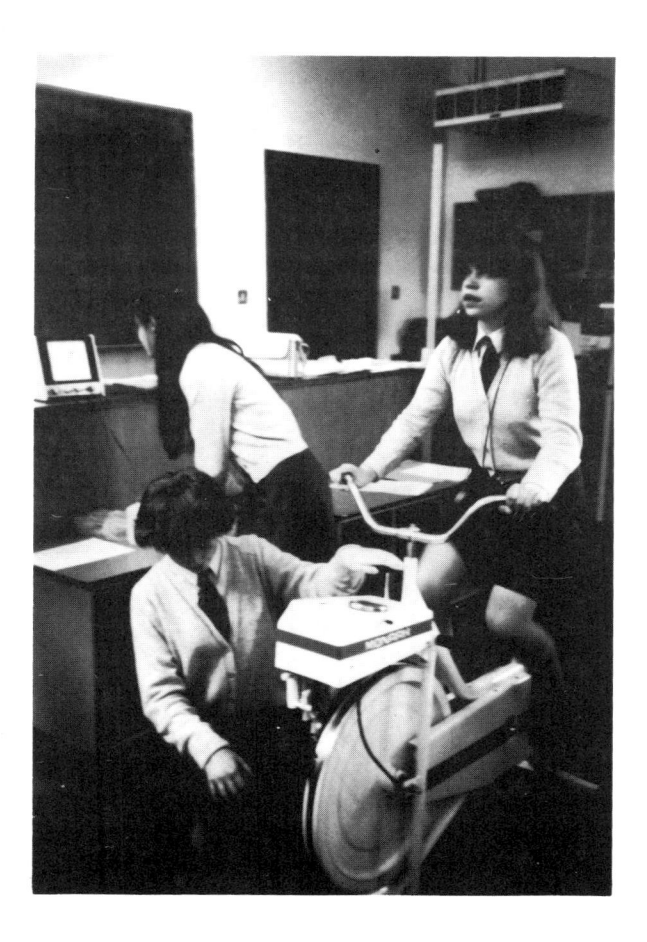

RESPIRATION AND VENTILATION

Examples suitable for Key Stage 3, range from comparing the amount of oxygen or carbon dioxide in atmospheric (inhaled) air) and expired air (Mackean, 1975), to measuring lung capacity (Roberts and Mawby, 1991). Variables can be introduced for Key Stage 4 and higher, such as the change in vital capacity with chronological age or in girls with the onset of the menarche (Bainbridge, pers. comm.).

The relationship of breathing and heart rates to exercise can be investigated using a cycle ergometer, and the effects of changing air composition on breathing rates, together with measurements of oxygen consumption, using a spirometer (Nuffield Biology, 1985). A spirometer must be use safely and guidance should be sought before starting work, e.g. that found in CLEAPSS (1991). The spirometer normally connects physically to a typograph or chart recorder which records rates and depths of an individual's breathing movements. However, since the introduction of data-logging, a universal interface can be connected to a position sensor (Frost, 1993). This has a lever arm which can replace the chart recorder and provide a more useful graph. This exciting and user-friendly book contains a collection of suggested experiments using sensors which is suitable for pupils aged 11-18 years. The main drawback is the cost of the equipment, for each group of pupils will need access to a computer and a universal interface. In addition, a variety of sensors are needed, designed to record and display the readings of measurements of such parameters as sound, temperature, humidity, pressure, position and rotation of objects. For schools that have invested in the LogIT data logger from Griffin & George, it is possible to display and analyse breathing and heart rates, using heart and breathing monitor senses and associated software. Primary pupils can investigate lung capacity using such simple equipment as a large previously calibrated and inverted sweet jar and some rubber tubing (Ward, 1983). The results of the class can be put on to a large histogram and perhaps related to pupils' height (Health Education Council, 1990).

BLOOD CIRCULATION

Arterial blood pressure can be measured directly using a sphygmomanometer. Better still, using a universal interface and a pressure sensor attached to the cuff it is possible to obtain a picture of pressure changes as the cuff pressure is increased and then released (Frost, 1993). Pulse measurements can be measured directly by pupils working in pairs and variables can be introduced such as changing activities from lying down, sitting, standing and degrees of exercises. Again, using a pulse sensor attached to the ear lobe or the chest allows the pulse to be monitored graphically. Pupils can increase their understanding of the cardiovascular system by using the computer simulation program, Blood Circulation Maze (AVP, 1993).

One teacher records how he found children in junior classes could investigate heart beats using their Lunar Zoomers (purchased from Woolworth's), as stethoscopes (Ward, 1983).

FOOD AND DIET

The topics of food and diet provide useful sources of investigations which are directly related to ourselves. Experiments can cover topics such as food chemistry, dental hygiene, ergonomics, and the role of enzymes in the digestion of food.

Nutrition is a basic life process about which all pupils should understand something. Food tests provide a cheap and simple source of suitable class experiments which can be tackled at all levels. Primary teachers can find clear instructions (Mackean, Worsley and Worsley, 1982) for testing for starch, reducing sugars, fat and protein. A comprehensive set of packs on all aspects of nutrition is available from the Nutrition Foundation (1992, 1993). Separate packs are available for each Key Stage, 5-6 year olds, 7-11 year olds, 12-14 year olds and 15-16 year olds.

Investigations involving enzymes are useful activities for teaching and learning about how to conduct scientific investigation. Pupils can, for example, be encouraged to investigate the conditions which control the activity of salivary amylase on starch. The teacher can introduce the role of key variables such as temperature, pH or substrate concentrations, and ask pupils to investigate their effects on the rates of enzyme activity (Mackean, 1971).

Because saliva can spread infections such as colds and sore throats, proper hygiene must be observed. The use of human saliva provides a good opportunity in biology lessons for teaching about hygiene. Pupils should use only *their own* saliva samples *and*, at least at secondary level, should be responsible for cleaning up their own equipment. Glassware contaminated with saliva should be placed directly after use into a vessel already containing freshly-prepared hypochlorite disinfectant, before it is washed with hot water and detergent. Tables and benches should be wiped with disinfectant, and hands washed.

HEALTH EDUCATION

Health and social education feature prominently in biology and can be promoted by investigations into the dangers of smoking, drug and alcohol addiction, and the effects of diet on health, body weight and dental decay. Some of these investigations can profitably involve using pupils as a resource.

Teeth and the buccal cavity provide an interesting area for study. The change in acidity of the mouth can be investigated before eating and at fixed times after by testing its pH (Burns, personal communication). The rate at which acid is produced in the mouth during eating has been shown to depend on the numbers of *Streptococcus mutans* present in dental plaque (Handley, 1992). Many pupils also enjoy using disclosing tablets (containing erythrosin) to reveal deposits of plaque, and they can subsequently examine the effect of cleaning teeth, with and without toothpaste, on its distribution.

Human energy requirements provide a rich seam for experimentation. There are several useful computer programs that enable pupils to use data about themselves. Microdiet (Philip Harris) and Diet Analysis (Longman AVP) enable pupils to find the nutrients present in foods, while Human Energy Expenditure (Longman AVP) calculates the energy required for different activities and the energy obtained from different foods.

The subject of drugs is not an easy one to investigate experimentally, but in considering the effects of drugs, aspirin (salicylic acid) can be shown to inhibit the activity of salivary amylase (Freeland, 1987). The damage caused by smoking can be investigated by setting a model cigarette-smoking machine (Morgan and Robson, 1982) and collecting the gaseous products by passing them through limewater, universal indicator and cotton wool, to demonstrate their acidity and the production of a tarry residue.

BEHAVIOUR AND RESPONSE

Human behaviour provides a variety of patterns which pupils can investigate, from simple observations on the sense organs, measuring rates of response to a variety of stimuli, to studying modes of learning.

Much work in science at primary school level is based around investigations of the senses, and there are many simple observations that pupils can make to test senses as, for example, described by Mackean (1978), Freeland (1987) and Ward (1983).

Variations can be introduced to extend the work, for example, finding out the least concentration of sugar or salt which a pupil can detect by taste.

There is a variety of simple investigations that can be carried out to investigate how the eye 'sees', and the ear 'hears', such as blindfolding a subject, then seeking out a ticking object or determining the distance over which a watch can be heard (Roberts, 1986; Ward, 1983. p 27).

Finding out about the sensitivity of the skin to touch lends itself to open-ended investigations. Using a blunt pencil, one can compare the sensitivity of the finger tips with other areas of the body, or investigate the sensitivity of the skin by tests with objects at different temperature or varying the delicacy with which it is touched (Mackean, 1978). Pupils can, when blind-folded, compare their ability to recognise by touch a variety of textured materials.

Pupils of all ages enjoy seeing how quickly they react to stimuli. A number of software packages investigate whether one reacts more quickly to sounds or to visual stimuli. In the absence of any computer software or other equipment to quantify reaction times, a good index of speed of reaction can be obtained by measuring how far a metre rule, held by pupil A, falls before being grasped by pupil B. Pupil B stands with his/her fingers almost touching each side of the metre rule and is not, of course, aware of when pupil A will release the ruler.

Finally, the whole field of optical illusions is a fascinating one. Optical illusions show how the human brain does *not* accurately read what it sees. Rather, it *interprets* what it sees.

The influence of experience on learning provides an interesting investigation for pupils. This can be done using mirror drawing where the hand cannot be seen as the writer is timed in an attempt to trace around outlines of stars, with first the writing hand and then the other (Mackean, 1978). Right-handed pupils may be surprised to find learning the task using the left hand may be faster than with their right hand. A useful collection of four books for practical investigations of the senses is available from the Association for Science Education (CLEAPSS, 1994).

GENETICS

Humans are difficult to study as controlled breeding experiments are obviously not possible, however, one can establish a genetic connection by looking for evidence of patterns in family trees and seeing if they conform to Mendelian principles.

There are several easy exercises for primary pupils, for example, they can clasp hands and find out whether they are left-thumbed (with left thumb on top) or right-thumbed. This trait should run in families. They can also have fun investigating variation in fingerprints using talcum powder, and then pressing the middle fingertip on to transparent adhesive tape, and finally sticking the 'print' on to black paper (Ward, 1983. pp 76 - 77).

Many secondary students find the study of genetics quite difficult, but discussion of examples of human genes which behave in a Mendelian fashion certainly helps the subject to become more comprehensible.

The classic example of a human character being determined by a pair of alleles at a simple gene is eye colour - with brown being dominant to blue. Pupils can be encouraged to investigate the inheritance of eye colour within their families. However, caution should be taken for two reasons. First, the brown allele is *not* always dominant to the blue allele (Reiss, 1988). Secondly, not all children live with their biological parents.

Tongue rolling is often described in school textbooks as a genetically-determined trait. In fact most non-rollers can be taught to roll their tongues (Patefield and Moore, 1986). This strongly suggests that tongue rolling is not simply genetically determined. The ability to taste phenyl thio carbamide (PTC) which can be performed safely using prepared paper strips available from Philip Harris does, however, show a simple pattern of inheritance.

If a set of colour blindness test cards and enough pupils are available, it can soon be shown that red-green colourblindness is a son-linked character. About 8% of males, but only 1% of females, are red-green colourblind.

Humans also provide good evidence of multiple genetic (polygenetic) effects such as variation of height or mass in sample groups. Any such observed variable will allow the teacher to introduce the idea of the importance of environmental effects as well as the influence of genes on phenotypic characters (Reiss, 1987).

GUIDELINES AND ETHICAL PRINCIPLES

Teachers should always ensure that the experiments they carry out conform to acceptable standards of safety. Not only this but where some behavioural or physiological aspect of a human is the subject of the investigation, great care must be taken to conform to acceptable ethical guidelines. The British Psychological Society has recently revised its ethical principles for all psychologists conducting research with human participants (BPS, 1993). It has advised members to draw these principles to the attention of other colleagues working in the field of human research.

All teachers who supervise pupils conducting investigations which contain an element of human biology should ensure that they are familiar with these guidelines. This is advisable both for the safety of pupils and for the teacher's own protection from the law. In recent years there has been an increase in legal actions by parents against teachers for alleged misconduct. Teachers should be aware of the possibility of such legal action if they infringe on the rights and dignity of pupils involved in investigations. Pupils come from a variety of cultural and ethnic origins and they will no doubt be of different ages, gender and social backgrounds. The teacher may have insufficient knowledge of each participant to realise the psychological implications of the results of investigations on some pupils.

The following list of amended points has been selected for the benefit of teachers from a checklist (see Figure 5·2) drawn up by the British Psychological Society (BPS, 1992) for research workers.

1. Have you tried the procedure on yourself?
2. Have you done a pilot study?
3. Are participants likely to experience discomfort or be offended or embarrassed in any way?

4. Where practical, if using persons under sixteen years of age, have you obtained the consent of parents or those in loco parentis?
5. Have pupils been kept fully informed as to what they are being asked to do and why?
6. Will pupils be told the results after the study?
7. Will the results be kept confidential, after the study is complete?
8. Have pupils been told that they may withdraw from participation if they want to at any time without any form of penalty?

In short, it is important that a teacher plans out any proposed study with early and careful preparation. If a teacher has any doubts about the safety or efficacy of the procedure, it is best not to go ahead with it or to consider if there are other more suitable ways of carrying out the investigation.

In addition to teachers using human subjects in class work, observational studies are likely to be quite attractive to advanced level students as an extended investigation or project work, because they are relatively easy to set up. The teacher in charge should decide whether any proposed project provides a suitable topic for research and falls within the ethical guidelines above. In addition, in such studies where the study is likely to be presented to an examiner or even published, care must be taken that any results obtained from participants are not identifiable.

Figure 5·2. Investigation Checklist

The Basic Rules for Ethical Psychological Research

Reproduced by kind permission of Dr Lesley Cooke and Angela Hughes, Chester College.

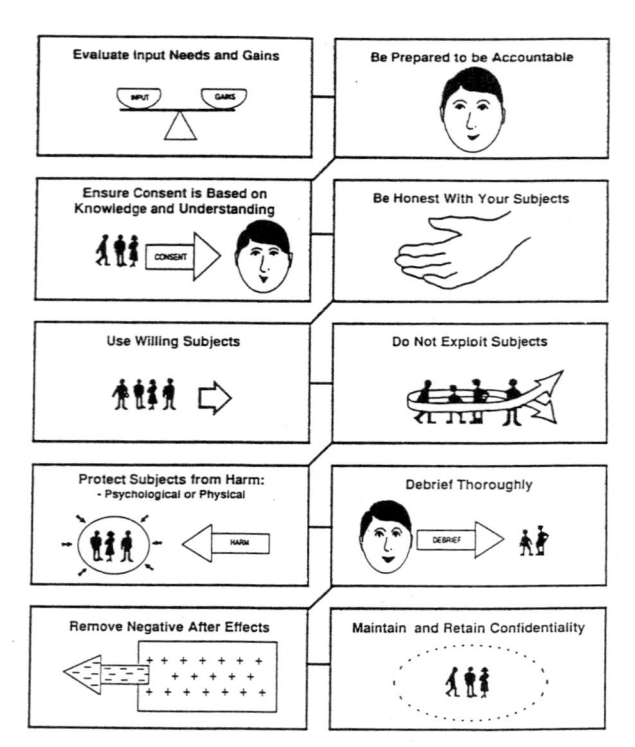

- Evaluate Input Needs and Gains
- Be Prepared to be Accountable
- Ensure Consent is Based on Knowledge and Understanding
- Be Honest With Your Subjects
- Use Willing Subjects
- Do Not Exploit Subjects
- Protect Subjects from Harm: - Psychological or Physical
- Debrief Thoroughly
- Remove Negative After Effects
- Maintain and Retain Confidentiality

REFERENCES

BNF (1992, 1993) *Food - A Fact of Life*. London: British Nutritional Foundation.

BPS (1993) *Ethical Principles for Conducting Research with Human Participants*. Leicester: British Psychological Society.

BPS (1992) *Manual of Psychology Practicals. Experimentation, Observation and Correlation*. Leicester: British Psychological Society. pp 207 - 210.

CLEAPSS (1991) *Laboratory Handbook*. London: Brunel University. p 1435. (Available only to members of CLEAPSS).

CLEAPSS (1993) *Equipment & Materials for Human Physiology R200a 7 b*. London: Brunel University (Available only to members of CLEAPSS)

CLEAPSS (1994) *The senses: Introducing the senses, Taste, Smell & Touch, Hearing & Balance, Eyes & Vision*. Hatfield, Herts: Association for Science Education.

DES, WOED (1987) *AIDS: Some Questions and Answers*. London: HMSO.

Encyclopaedia Britannica (1992) *Science and Technology Video Series. No 10, The Digestive System*. Wallington, Surrey: Encyclopaedia Britannica International Ltd.

Freeland P.W. (1987) *Investigations for GCSE Biology*. Sevenoaks: Hodder and Stoughton.

Frost R. (1993) *Datalogging and Control*. Hatfield, Herts: Association for Science Education.

Handley P. (1992) Streptococcus mutans and Tooth Decay. *Biological Sciences*, **5**(2), pp 12 - 14.

Health Education Council (in association with Sheffield Education Department) (1990) *My Body Project*. London: Wayland.

IOB (1987) *Some Recommendations on Blood Sampling*. London: Institute of Biology.

Langley G.R. (1991) Animals in science education - ethics and alternatives. *Journal of Biological Education*, **25**(4), pp 274 - 279.

Mackean D.G. (1971) *Enzymes, Experimental Work in Biology, 2*. London: John Murray.

Mackean D.G. (1975) *Respiration and Gaseous Exchange, Experimental Work in Biology, 7*. London: John Murray. pp 6 - 9.

Mackean D.G. (1978) *Human senses, Experimental Work in Biology, 8*. London: John Murray.

Mackean D.G., Worsley C.J. and Worsley P.C.C. (1982) *Class Experiments in Biology*. London: John Murray. pp 17 - 24.

Mansell J. (1989a) Two helpful models for teaching about blood. *Journal of Biological Education*, **23**(3), p 179.

Mansell J. (1989b) A 'model' blood group. *Journal of Biological Education*, **23**(1), pp 4 - 5.

Morgan A.G and Robinson M. (1982) *Introducing Biology*. London: Macmillan Education. p 104, p 165.

Nuffield Advanced Level Biology (1985) *Practical Guide 1: Gas Exchange & Transport in Plants and Animals*. Harlow, Essex: Longman.

Patefield J.M. and Moore M.L. (1986) The genetic basis of tongue rolling. *Journal of Biological Education*, **20**(4), pp 255 - 266.

Reiss M.J. (1987) The inheritance of height: environmental influence and polygenic effects. *Journal of Biological Education*, **21**, p 159.

Reiss M.J. (1988) Brown-eyed children may have blue-eyed parents. *School Science Review*, **69**, p 742.

Roberts M.B.V. (1986) *Biology for Life*. London: Nelson. p 297.

Roberts M.B.V. and Mawby P.J. (1991) *Biology*. Hong Kong: Longman Group.

Rogers W.G. (1985) Computer assisted learning programmes and class management. In: *Exploring Biology with Microcomputers, Readers* (4) (ed. Smith C.J.). London: Council for Educational Technology (on behalf of the Microelectronics Educational Programme). pp 54 - 62.

Tomlins B. (1988) *Guidance on the use of Living Material in the Classroom*. London: Institute of Biology.

Ward A. (1983) *A Source Book of Primary Science Education*. Sevenoaks: Hodder and Stoughton.

Wells J. (1991) Observing cells in plucked hair follicles. *Journal of Biological Education*, **25**(1), pp 3 - 4.

Chapter 6 LEGALITIES

C. Bywater

ABSTRACT

There is an overriding principle that the treatment of animals in society must avoid any unnecessary suffering. Behind this principle, however, lies a mass of often complex legislation which may inhibit, or even prohibit, a number of activities which teachers might wish to do in the interests of education.

These could range from a simple desire to look after an injured animal or wild bird brought to school by a pupil, or an analysis of either animal tissue or plant membrane at a secondary level.

Any of these activities may, in fact, be regulated or otherwise authorised only in Universities and/or Colleges of Higher Education. Equally a large variety of animals and a whole host of plants are protected in a variety of ways.

Accordingly, whilst there are certain basic standards concerning the general welfare of animals, health and safety implications and old Common Law obligations, which should be observed, it is also important for teachers to have some indication of other specific Acts of Parliament or subordinate regulations made by Statutory Instrument through the appropriate Government Department which might apply.

Many may not, in fact, be relevant in most circumstances, and common sense will often be sufficient, but this will not always be the case. This paper, therefore, sets out to explore the general principles, to identify relevant legislation which may require consideration, and to provide overall guidance and advice, together with information on where to make further enquiry in specific cases.

INTRODUCTION

In today's society, perhaps more than ever, there are sound educational reasons for the study of animals and plants in schools.

This undoubtedly stems from the gradual development of philosophy and general analysis about the nature of our environment. Whilst, historically, plants were used for their herbal and medicinal properties, wood for houses and ships of the line, and even charcoal for iron, animals, rather surprisingly, did not always fare so well. Although the work of Linnaeus and the art of Dürer and Stubbs were evidence of a new enlightenment, many animals were not well treated.

However, the early beginnings of reform in the last century resulted in the Ill Treatment of Cattle Act of 1822. Shortly thereafter, William Wilberforce became co-founder in 1824 of the RSPCA, which was designed to enforce the new law. This Act was subsequently replaced by the Protection of Animals Act 1911 (the "1911 Act") though an intervening Act, the Cruelty to Animals Act 1876, was also replaced by the Animals (Scientific Procedures) Act 1986, referred to below.

The 1911 Act remains the most important law in relation to the treatment of animals in schools. By its reference to ill-treatment in general, including the fighting or baiting of animals, on the one hand, and the administration of any poisonous or injurious drug or substance on the other, it provides an area of agreement between the old and the modern scientific era.

Essentially, the 1911 Act prohibits certain acts of cruelty, to include abuse, neglect or causing any unnecessary suffering to captive and domestic animals, including birds, fish and reptiles. Penalties, apart from a fine, include imprisonment, which at the time could have involved hard labour, though the Act has been amended a number of times since then. The definition of "animal" includes the expression "of whatsoever kind of species". Thus, invertebrates might be included, though the examples given of both domestic or captive animals do not suggest that the draughtsmen of that Act had them in mind.

Since the 1911 Act does not specifically mention schools, the use and care of animals is subject to the same laws which apply to the general public in an attempt to prevent illtreatment and unnecessary suffering wherever that may occur.

In that case, one might reasonably ask what is and what is not permitted in a science or biology laboratory at school? Unfortunately, as with most law, there is no simple answer. In fact, there are a significant number of other laws which may have a bearing. However, since it would take a book, rather than a chapter, to cover these in detail, I have chosen to divide this subject into those topics or activities which seem most relevant, together with an indication of the applicable law.

I have included appropriate details of those animals and plants which are currently protected under various Acts or subordinate regulations (Statutory Instruments), together with certain scientific procedures which are unlikely to be permitted in schools because they would be regulated under the Animals (Scientific Procedures) Act 1986, described below.

GENERAL WELFARE

It will be apparent from the above that the maintenance and general welfare of animals in schools must avoid any suffering, distress or neglect, and this, obviously, applied to weekends and holidays when proper provision must also be made. Care should, therefore, be taken to ensure that animals are well fed and watered on a daily basis and (so far as necessary) visited at suitable intervals. Cages and aquaria should be cleaned and properly maintained and be otherwise suitable in terms of size, temperature, lighting and ventilation.

These parameters are contained in the Pet Animals Act 1951 (the "1951 Act") which may not, in fact, apply, but guidance can be obtained from other statutes in giving further meaning to the expression "unnecessary suffering" in both the 1911 Act and the Abandonment of Animals Act 1960 (the "1960 Act"), described below.

Indeed, both the 1951 Act and the 1960 Act do have implications concerning the disposal of surplus animals. A pet shop is very broadly defined so animals should not advisedly be sold to pupils, and it is an offence to sell pets to children under twelve, nor should mammals be sold at too early an age. If in doubt whether the 1951 Act applies, the Department for Education's advice (1990) is to obtain confirmation from the local authority responsible for licensing.

That advice also recommends that schools should always involve parents, even where animals are given to pupils, since distress and suffering may occur as a result of subsequent abandonment, which would therefore be an offence under the 1960 Act. Clearly, indigenous non-domestic fish and amphibians, or invertebrates, can be returned to their natural habitat.

Apart from the above, there are a number of Acts or subordinate regulations which have implications concerning the transport and supply of animals, though some of these are primarily the concern of commercial suppliers and LEAs which operate loan or supply systems. They deal with protection against injury or unnecessary suffering of animals during loading, unloading or carriage, and with feeding, watering and general care during that process. If any school is concerned with transport or supply they should seek advice from the local school's authority or the Department for Education (DFE).

Finally, there is sound advice in other chapters of this book concerning the maintenance and well-being of a whole variety of vertebrates. Provided such standards are maintained there should be no cause for concern.

HEALTH AND SAFETY ASPECTS

The Health and Safety at Work etc. Act 1974 has certain implications for work in schools. Animals from reliable sources should not cause problems if teachers take normal care and proper hygienic precautions in handling them. However, the possibility of certain transmissible diseases and allergic reactions should be considered. Advice here can be obtained from the Health and Safety Executive, Broad Lane, Sheffield, S3 7HQ (telephone 01742 892000).

It is also illegal, without a licence, to keep certain scheduled wild animals, e.g. non-domestic canines, wild cats, most monkeys and apes, crocodiles, alligators and poisonous snakes, under the Dangerous Wild Animals Act 1976. The Animal Health Act 1981 is, perhaps, important for educational purposes dealing, as it does, with quarantine for animals susceptible to rabies. It is also concerned to prevent the spread of other serious animal diseases, such as swine fever, foot and mouth and other cattle diseases such as bovine spongiform encephalopathy, or BSE.

COMMON LAW

Teachers should also be aware that each member of society owes a general duty of care at law to refrain from any act, omission or neglect which is capable of causing harm to those who might forseeably be affected. Teachers, being in 'loco parentis', clearly have that responsibility toward their pupils, though the duty is broader than that and could apply to anyone who might be harmed, to include other staff or parents.

This has nothing to do with any Act of Parliament, but is part of our old Common Law obligations, in particular, the civil law of 'tort', which dates back to the Norman era, hence the word's French origin. These laws apply to all of us in our day to day activities and are, perhaps, best exemplified by the care we take when driving. An accident could well result in a criminal prosecution under the Road Traffic Act leading to the imposition of a fine or penalty. It could equally result in civil proceedings against the individual personally for financial compensation, which might greatly exceed any such fine. Most of the Acts mentioned in this article are actually part of the criminal law and do not essentially cover civil claims (though it can be possible to sue for breach of statutory duty). Although one may be insured, and teachers will no doubt be covered by their employers' liability insurance (except, perhaps, in cases of extreme recklessness or wilful behaviour), such claims are best avoided. It is prudent to sound a cautionary note, as we are now in a more litigious society, evidenced, perhaps, by the willingness of children to sue their parents under the Children Act 1989.

80 Chapter 6 Legalities

Fortunately, Common Law is largely common sense, so it makes sense not, at least, without a trained keeper, to bring dangerous animals to school or do anything which might cause injury, or even trauma. Equally, the chapter on Pupils as a Resource provides good guidelines and ethical principles in that connection which should be observed.

PERMITTED PROCEDURES

The title of this section is slightly misleading since most laws tend to be negative rather than positive in the sense that they state what is not permitted rather than that which is. England and Wales do not have a detailed 'codified' legal system and the laws are often expressed in fairly broad, but complex, terms, leaving it up to the Courts to determine their precise meaning in cases of doubt. Past precedent and expert evidence will usually be relied on.

An example of this is the Animals Scientific Procedures Act 1986 (the "1986 Act") which replaced the Cruelty to Animals Act 1876. The 1986 Act institutes a system of personal and project licences in designated establishments for the carrying out of regulated procedures as defined below. Since a project licence will only be given for education and training in designated establishments "otherwise than in primary or secondary schools" it seems unlikely that one will be given for any of the other research purposes specified.

However the 1986 Act still applies to schools, in the sense that, if a regulated procedure is carried out, it would be illegal. By contrast, although the 1986 Act is unlikely to apply in any other way to schools, it is still advisable to adopt the Standard Methods of Human Killing contained in Schedule 1 of the Act to avoid any implication of cruelty or unnecessary suffering under the 1911 Act. Schedule 1 to the 1986 Act is, therefore, reproduced at the end of this chapter.

These methods should be employed if it is necessary to kill animals, not otherwise protected, for dissection or for physiological work on tissues or organs, since this can only be a regulated procedure in designated establishments where such methods are not employed. However, it is strongly urged by the RSPCA and others that the killing of animals should not occur as part of a lesson nor, indeed, take place in front of pupils. This is arguably good advice on Common Law grounds and it may well be preferable to obtain dead animals from licensed suppliers. In this connection, the 1986 Act also controls establishments for the breeding and supply of animals which schools might wish to use, although they have no obligation to do so, as these are designed for use in regulated procedures.

A "regulated procedure" under the 1986 Act means any activity. other than humane killing, which is likely to cause a "protected animal" (as defined) any "pain, suffering, distress, or lasting harm" which, as we have seen, is only likely to be a licensed activity in universities and/or colleges of higher education.

The definition of protected animals is any living vertebrate "other than Man" (who, though excluded from the Act, is otherwise protected!), and any such vertebrate in its foetal, larval or embryonic form, but only from the stage of its development when:

    a)   in the case of a mammal, bird or reptile, half the gestation or incubation period for the relevant species has elapsed; and

    b)   in any other case, it becomes capable of independent feeding.

It should, perhaps, be noted that "vertebrate" is here defined as any animal of the Sub-phylum Vertebrata of the Phylum Chordata and "invertebrate" means any animal not of that Sub-phylum. The Act does not at present affect invertebrates, but the Home Secretary is empowered to include them by Statutory Instrument if necessary.

Any such vertebrate or animal shall be regarded as continuing to live until the permanent cessation of circulation or the destruction of its brain.

The administration of anaesthetics and analgesics, together with decerebration, are also regulated procedures which cannot, therefore, take place in schools.

According to the DFE's advice (1990) it follows from these points that the following practices, carried out on unprotected animals as defined in the 1986 Act, are illegal in schools:

a) decerebration of an animal by pithing
b) the carrying out of the gastrocnemius nerve/muscle preparation on amphibians, unless carried out after humane killing
c) the injection of hormones to cause spawning in *Xenopus* (African clawed toad), unless this is carried out solely for husbandry purpose
d) nutritional experiments involving restricted or excessive diets, although the usual investigation of plotting the growth rate of a mammal on a normal diet is perfectly acceptable
e) an experiment involving the tossing of a young mouse from hand to hand
f) anaesthetizing an animal (except for humane killing by administration of a lethal overdose)
g) any regulated procedure on a tadpole once it is capable of independent feeding; although a tadpole could be expected to have reached this stage two days after hatching, teachers should note that this could be earlier or later, depending on variations in temperature
h) the examination of living chick embryos after the halfway point of incubation.

It should be apparent from what has been said above that if an animal is protected by the 1986 Act and the procedure proposed is regulated, then it cannot be carried out in a school.

Of course, animals may clearly be used for observation, drawings and procedures which are not regulated under the 1986 Act or otherwise. In this respect it has been suggested by O'Donoghue (1988) that the best guide is not to depart from what is normal for the particular animal. For example, a rodent may well voluntarily wish to use an activity wheel or pursue certain pathways in a maze. Other procedures may be carried out where they are within the normal range experienced by the animal or are part of a natural process.

Lock (1989) has suggested that, whilst teachers would clearly wish to avoid pain, suffering, distress or lasting harm to an animal, the signs of distress caused by behavioral investigations may not be clear cut, particularly where the level of distress is low and indistinguishable from normality. In investigating the development of reflex actions in young mice it would be difficult to determine whether or not the animals were distressed.

It is probably safe to assume that any behavioral investigations in the case of vertebrates within the ordinary limits of *e.g.* temperature and light experienced by the animal are permitted. It would also be acceptable to investigate food preference and feeding behaviour provided that this avoids restricted or excessive diets and that any variation in established dietary ingredients could also be experienced in the animal's natural habitat (Lock, 1989).

Finally, there is no reason why schools cannot breed animals on the premises though, clearly, the reproduction of vertebrates with inherited defects should be avoided. Not only might this cause distress, it could represent a regulated procedure under the 1986 Act.

OTHER PROTECTED SPECIES

Apart from the protection provided for humane purposes under the 1986 Act, a whole variety of animal and plant species are protected under various Acts. Many of these are wholly unlikely to come anywhere near schools for the purposes of this paper, since we are not here concerned with field studies.

However, this is not impossible, especially in rural areas, and the following should therefore be noted:

1. Deer, seals and badgers are protected, so schools should not accept them unless they were obtained otherwise than in contravention of the relevant Acts (Deer Act 1991, Conservation of Seals Act 1970, Protection of Badgers Act 1992).
2. Imports and exports of various animals are prohibited without a licence. Any school wishing to do so should refer to the Department of the Environment (DOE) (Wildlife Division), (Endangered Species (Import and Export) Act 1976). (For non-domesticated/non-native birds, mammals and reptiles, contact CITES branch, DOE, Bristol on 01272 218168 (General Enquiries).
For Wildlife Trade Licensing contact DOE, Bristol on 01272 218694, for birds, or 01272 218291 for other animals and plants.)
3. Under the Wildlife and Countryside Act 1981 (the "1981 Act") a whole variety of other animals are also protected. Since we are not concerned with fieldwork, salient points only will be made for educational purposes generally and for those schools whose grounds may contain a variety of fauna and flora.

In the first place, all wild birds and their eggs and nests (when in use) are protected to some degree, except pest species, which can only be killed by authorised persons (including the landowner), and certain game birds. Some birds and their nests may not even be approached or photographed. The use of wild birds is strictly controlled and certain birds require to be registered and/or ringed if kept in captivity.

Should any teacher or pupil wish to bring a wild bird to school, advice should be sought, in the first instance, from the DOE, Bristol on 01272 218693, since the Act is highly complex. It is, for example, possible to look after a bird that has been accidentally disabled until it recovers though it should be reported and may need to be registered. Equally, how it is confined and any other activity relating to wild birds is likely to be regulated.

The Royal Society for the Protection of Birds (RSPB) discourages the use of captive birds for educational purposes largely on two counts. In the first place several speakers have been prosecuted for infringements of the 1981 Act, especially in relation to owls and raptors. They are also concerned that young people may be stimulated to possess birds for themselves, their obsession forcing them to break the law by taking birds from the wild, and this will frequently result in cruelty to such birds.

RSPB believes there are many alternative ways of interesting and educating children in the conservation of wild birds and, whilst this is opinion and not law, though it is designed to prevent illegal activity, their view has been adopted by a number of LEAs. In this connection it might be noted, as a general point, that schools under LEA control are bound by rules laid down by the LEA as the employer.

The RSPB has also endorsed RSPCA recommendations in their "Guide to Visiting Animal Schemes" which includes a 'code of conduct' and a checklist of questions for teachers to consider before a speaker is invited to a school. It goes without saying that the RSPB also supports the use of photographs and other audio-visual aids, which correspond more appropriately to modern educational methods, rather than the use of dead/parts of birds, eggs and nests. Again this is pre-emptive advice, since proof of legal possession may be difficult and the discouragement of illegal collection paramount on ecological grounds.

Many other animals are also protected by the 1981 Act in a similar way. However, it seems appropriate to append a list of these animals to this article, since some of them might conceivably be brought to school, *e.g.* certain moths and a variety of butterflies, though walruses, also included, might prove a little less accommodating!

The list also contains an indication of the degree of protection. For example some reptiles and amphibians (including their spawn and tadpoles) have full protection, whereas the common frog is only partially protected. Fully protected species must not be brought into school or removed from their natural habitat.

For the rest, only a very small number should be brought into school and these should be returned as soon as possible, preferably to the place where they were collected. It is most important, for example, that reptiles and amphibians and hibernating mammals should be released in time to build up the fat stores essential for their hibernation. Apart from ethical and ecological considerations, any other action could cause unnecessary suffering, distress or lasting harm.

If pupils bring protected species inadvertently to school, the opportunity should be taken to explain why they are being returned to their habitat as well as the fact that it is illegal to be in possession of protected species. Further information is available in the DOE booklet "Protecting Britain's Wildlife" which is available from local authorities.

It should also be noted that it is an offence to damage, destroy or obstruct access to any structure or place used for shelter or protection by a wild animal, or to disturb such an animal in its shelter. The 1981 Act also limits the methods by which some of the animals listed below may be killed or taken.

Finally, the 1981 Act also has considerable implications in relation to plants. Unless the school has a licence from the DOE, it is illegal intentionally to pick, uproot or destroy any of the specially protected plants, also listed below, or to uproot any plant grown in the wild (unless with the permission of the landowner). Schools should therefore take careful note of the need to obtain supplies from private sources if they do not have a licence from the DOE.

APPLICABLE LAW

The law expressed in this paper is that relating to England and Wales only. (Some may relate in whole or part to Scotland and Northern Ireland, but there are considerable variations in both cases.)

Whilst the basic standards required in the treatment of animals are likely to remain the same, statutory and regulatory controls are subject to amendment or variation from time to time. For example, an Act relating to one or more specific types of animal, such as the badger, might be consolidated or otherwise brought up to date as has happened since the DFE publication (1990).

Equally, it should be noted that the lists of animals and plants have also been considerably extended by Statutory Instrument since that publication, and this could recur in future. In cases of doubt it is better to enquire. Government Departments are usually very helpful, though patience may be required to find the right department. For example, telephone numbers given for the DOE will hopefully provide a useful starting point for any relevant enquiry.

STANDARD METHODS OF HUMANE KILLING

SCHEDULE 1

ANIMALS (SCIENTIFIC PROCEDURES) ACT 1986

Method	Animals for which appropriate
1. Overdose of anaesthetic suitable for the species -	
A. Animals other than Foetal, larval and embryonic	
(i) by injection	(i) All animals
(ii) by inhalation	(ii) All animals up to 1 kg bodyweight except reptiles, diving birds and diving mammals
(iii) by immersion	(iii) Fishes Amphibia up to 250 g bodyweight

(Followed by destruction of the brain in cold-blooded vertebrates and by exsanguination or by dislocation of the neck in warm blooded vertebrates except where rigor mortis has been confirmed).

2. Dislocation of the neck (Followed by destruction of the brain in fishes)	Rodents up to 500g bodyweight other than guinea-pigs Guinea-pigs and lagomorphs up to 1 kg bodyweight Birds up to 3 kg bodyweight Fishes up to 250g bodyweight Rodents up to 1 kg bodyweight
3. Concussion by striking the back of the head	Birds up to 250g bodyweight Fishes

Followed by exsanguination or dislocation of the neck in rodents and birds and destruction of the brain in fishes.

4. Decapitation followed by destruction of the brain	Cold-blooded vertebrates
5. Exposure to carbon dioxide in a rising concentration using a suitable technique followed by exsanguination or by dislocation of the neck except where rigor mortis has been confirmed	Rodents over 10 days of age up to 1.5 kg bodyweight Birds over 1 week of age up to 3 kg bodyweight

B. Foetal, larval and embryonic forms

1. Overdose of anaesthetic suitable for the species -	
(i) by injection	(i) All animals
(ii) by immersion	(ii) Fishes
2. Decapitation	Mammals

ANIMALS AND PLANTS WHICH ARE PROTECTED

(ORIGINAL SOURCE: WILDLIFE AND COUNTRYSIDE ACT 1981. Schedules 5, 6 &

8. However these have since been extensively amended by Statutory Instrument, and this could recur, so advice should be sought from either DOE or DFE.)

I ANIMALS

All these animals are fully protected apart from:

      *   species for which the offence relates to 'sale' only.
     **   species for which the offence relates to 'taking' and 'sale' only.
    ***   species for which the offence relates to 'killing, injuring and sale'.

Adder*	*Vipera berus*
Anemone, Ivell's Sea	*Edwardsia ivelli*
Anemone, Startlet Sea	*Nematosella vectensis*
Apus	*Triops cancriformis*
Bats, Horseshoe (all species)	*Rhinolophidae*
Bats, Typical (all species)	*Vespertilionidae*
Beetle, Rainbow Leaf	*Chrysolina cerealis*
Beetle, Violet Click	*Limoniscus violaceus*
Beetle	*Graphoderus zonatus*
Beetle	*Ilypebaeus flavipes*
Beetle	*Paracymus aeneus*
Beetle, Lesser Silver Water	*Hydrochara caraboides*
Beetle, Mire Pill (limited protection only)	*Curimopsis nigrita*
Burbot	*Lota lota*
Butterfly, Heath Fritillary	*Mellicta athalia* (otherwise known as *Melitaea athalia*)
Butterfly, Large Blue	*Maculinea arion*
Butterfly, Swallowtail	*Pipilio machaon*
Butterfly, Northern Brown Argus	*Aricia artaxerxes*
Butterfly, Adonis Blue	*Lysandra bellargus*
Butterfly, Chalkhill Blue	*Lysandra coridon*
Butterfly, Silver-studded Blue	*Plebejus argus*
Butterfly, Small Blue	*Cupido minimus*
Butterfly, Large Copper	*Lycaena dispar*
Butterfly, Purple Emperor	*Apatura iris*
Butterfly, Duke of Burgundy Fritillary	*Hamearis lucina*
Butterfly, Glanville Fritillary	*Melitaea cinxia*
Butterfly, High Brown Fritillary	*Argynnis adippe*

Butterfly, Marsh Fritillary	*Eurodryas aurinia*
Butterfly, Pearl-bordered Fritillary	*Boloria euphrosyne*
Butterfly, Black Hairstreak	*Strymonidia pruni*
Butterfly, Brown Hairstreak	*Thecla betulae*
Butterfly, White Letter Hairsreak	*Strymonidia w-album*
Butterfly, Large Heath	*Coenonympha tullia*
Butterfly, Mountain Ringlet	*Erebia epiphron*
Butterfly, Chequered Skipper	*Carterocephalus palaemon*
Butterfly, Lulworth Skipper	*Thymelicus acteon*
Butterfly, Silver Spotted Skipper	*Hesperia comma*
Butterfly, Large Tortoiseshell	*Nymphalis polychloros*
Butterfly, Wood White	*Leptidea sinapsis*
Cat, Wild	*Felis silvestris*
Cicada, New Forest	*Cicadetta montana*
Crayfish, Atlantic Stream**	*Austropotamobius pallipes*
Cricket, Field	*Gryllus campestris*
Cricket, Mole	*Gryllotalpa gryllotalpa*
Dolphins	*Cetacea*
Dormouse	*Muscardinus avellanarius*
Dragonfly, Norfolk Aeshna	*Aeshna isosceles*
Frog, Common*	*Rana temporaria*
Hatchet Shell, Northern	*Thyasira gouldi*
Grasshopper, Wart-biter	*Decticus verrucivorus*
Lagoon Snail	*Paludinella littorina*
Lagoon Snail, De Folin's	*Caecum armoricum*
Lagoon Worm, Tentacled	*Alkmaria romijni*
Leech, Medicinal	*Hirudo medicinalis*
Lizard, Sand	*Lacerta agilis*
Lizard, Viviparous***	*Lacerta vivipara*
Mat, Trembling Sea	*Victorella Pavida*
Marten, Pine	*Martes, martes*
Moth, Barberry Carpet	*Pareulype berberata*
Moth, Black-veined	*Siona lineata* (otherwise known as *Idaea lineata*)
Moth, Essex Emerald	*Thetidia smaragdaria*
Moth, New Forest Burnet	*Zygaena viciae*

Moth, Reddish Buff	*Acosmetia caliginosa*
Moth, Sussex Emerald	*Thalera fimbrialis*
Moth, Viper's Bugloss	*Hadena irregularis*
Newt, Great Crested (otherwise known as Warty newt)	*Triturus cristatus*
Newt, Palmate*	*Triturus helveticus*
Newt, Smooth	*Triturus vulgaris*
Otter, Common	*Lutra lutra*
Porpoises	*Cetacea*
Sandworm, Lagoon	*Amandia cirrhosa*
Sea Fan, Pink***	*Eunicella verrucosa*
Sea Slug, Lagoon	*Tenellia adspersa*
Shrimp, Fairy	*Chirocephalus diaphanus*
Shrimp, Lagoon Sand	*Gammarus insensibilis*
Slow-worm***	*Anguis fragilis*
Snail, Glutinous	*Myxas glutinosa*
Snail, Sandbowl	*Catinella arenaria*
Snake, Grass	*Natrix helvetica*
Snake, Smooth	*Coronella austriaca*
Spider, Fen Raft	*Dolomedes plantarius*
Spider, Ladybird	*Eresus niger*
Squirrel, Red	*Sciurus vulgaris*
Sturgeon	*Acipenser sturio*
Toad, Common*	*Bufo bufo*
Toad, Natterjack	*Bufo calamita*
Turtles, Marine (all species)	*Dermochelyidae* and *Cheloniidae*
Vendace	*Coregonus albula*
Walrus	*Odenbenus rosmarus*
Whitefish	*Coregonus lavaretus*

N.B. Adders are subject to the provisions of the Dangerous Wild Animals Act, 1976.

ANIMALS WHICH MAY NOT BE KILLED OR TAKEN BY CERTAIN METHODS

Badger	*Meles meles*

Bats, Horseshoe (all species)	*Rhinolophidae*
Bats, Typcial (all species)	*Vespertilionidae*
Cat, Wild	*Felis silvestris*
Dolphin, Bottle-nosed	*Tursiops truncatus* (otherwise known as *Tursiops tursio*)
Dolphin, Common	*Delphinus delphis*
Dormice (all species)	*Gliridae*
Hedgehog	*Erinaceus europaeus*
Marten, Pine	*Martes martes*
Otter, Common	*Lutra lutra*
Polecat	*Mustela putorius*
Porpoise, Harbour (otherwise known as Common porpoise)	*Phocaena phocaena*
Shrews (all species)	*Soricidae*
Squirrel, Red	*Sciurus vulgaris*

II PLANTS

Adder's-tongue, Least	*Ophioglossum lusitanicum*
Blackwort	*Southbya nigrella*
Alison, Small	*Alyssum alyssoides*
Broomrape, Bedstraw	*Orobanche caryophyllacea*
Broomrape, Oxtongue	*Orobanche loricata*
Broomrape, Thistle	*Orobanche reticulata*
Cabbage, Lundy	*Rhynchosinapis wrightii*
Calamint, Wood	*Calamintha sylvatica*
Caloplaca, Snow	*Caloplaca nivalis*
Catapyrenium, Tree	*Catapyrenium psoromoides*
Catchfly, Alpine	*Lychnis alpina*
Catillaria, Laurer's	*Catellaria laureri*
Centaury, Slender	*Centaurium tenuiflorum*
Cinquefoil, Rock	*Potentilla rupestris*
Cladonia, Upright Mountain	*Cladonia stricta*
Clary, Meadow	*Salvia pratensis*
Club-rush, Triangular	*Scirpus triquetrus*
Colt's-foot, Purple	*Homogyne alpina*

Cotoneaster, Wild	*Cotoneaster integerrimus*
Cottongrass, Slender	*Eriophorum gracile*
Cow-wheat, Field	*Melampyrum arvense*
Crocus, Sand	*Romulea columnae*
Crystalwort, Lizard	*Riccia bifurca*
Cudweed, Broad-leaved	*Filago pyramidata*
Cudweed, Jersey	*Gnaphalium luteoalbum*
Cudweed, Red-tipped	*Filago lutescens*
Diapensia	*Diapensia lapponica*
Dock, Shore	*Rumex rupestris*
Earwort, Marsh	*Jamesoniella undulifolia*
Eryngo, Field	*Eryngium campestre*
Fern, Dickie's Bladder	*Cystopteris dickieana*
Fern, Killarney	*Trichomanes speciosum*
Flapwort, Norfolk	*Leicolea rutheana*
Fleabane, Alpine	*Erigeron borealis*
Fleabane, Small	*Pulicaria vulgaris*
Frostwort, Pointed	*Gymnomitrion apiculatum*
Galingale, Brown	*Cyperus fuscus*
Gentian, Alpine	*Gentiana nivalis*
Gentian, Dune	*Gentianella uliginosa*
Gentian, Early	*Gentianella anglica*
Gentian, Fringed	*Gentianella ciliata*
Gentian, Spring	*Gentiana verna*
Germander, Cut-leaved	*Teucrium botrys*
Germander, Water	*Teucrium scordium*
Gladiolus, Wild	*Gladiolus illyricus*
Goosefoot, Stinking	*Chenopodium vulvaria*
Grass-poly	*Lythrum hyssopifolia*
Grimmia, Blunt-leaved	*Grimmia unicolor*
Gyalecta, Elm	*Gyalecta ulmi*
Hare's-ear, Sickle-leaved	*Bupleurum falcatum*
Hare's-ear, Small	*Bupleurum baldense*
Hawk's-beard, Stinking	*Crepis foetida*
Hawkweed, Northroe	*Hieracium northroense*

Hawkweed, Shetland	*Hieracium zetlandicum*
Hawkweed, Weak-leaved	*Hieracium attenuatifolium*
Heath, Blue	*Phyllodoce caerulea*
Helleborine, Red	*Cephalanthera rubra*
Helleborine, Young's	*Epipactis youngiana*
Horsetail, Branched	*Equisetum ramosissimum*
Hound's-tongue, Green	*Cynoglossum germanicum*
Knawel, Perennial	*Sceranthus perennis*
Knotgrass, Sea	*Polygonum maritimum*
Lady's-slipper	*Cypripedium calceolus*
Lecanactis, Churchyard	*Lecanactis hemisphaerica*
Lecanora, Tarn	*Lecanora archariana*
Lecidea, Copper	*Lecidea inops*
Leek, Round-headed	*Allium sphaerocephalon*
Lettuce, Least	*Lactuca saligna*
Lichen, Arctic Kidney	*Nephroma arcticum*
Lichen, Ciliate Strap	*Heterodermia leucomelos*
Lichen, Coralloid Rosette	*Heterodermia propagulifera*
Lichen, Ear-lobed Dog	*Peltigera lepidophora*
Lichen, Forked Hair	*Bryoria furcellata*
Lichen, Golden Hair	*Teloschistes flavicans*
Lichen, Orange Fruited Elm	*Caloplaca luteoalba*
Lichen, River Jelly	*Collema dichotomum*
Lichen, Scaly Breck	*Squamarina lentigera*
Lichen, Stary Breck	*Buellia asterella*
Lily, Snowdon	*Lloydia serotina*
Liverwort	*Petallophyllum ralfsi*
Liverwort, Lindenberg's Leafy	*Adelanthus lindenbergianus*
Marsh-mallow, Rough	*Althaea hirsuta*
Marshwort, Creeping	*Apium repens*
Milk-parsley, Cambridge	*Selinum carvifolia*
Moss	*Drepanocladius vernicosus*
Moss, Alpine Copper	*Mielichoferia mielichoferi*
Moss, Baltic Bog	*Sphagnum balticum*
Moss, Blue Dew	*Saelania glaucescens*

Moss, Blunt-leaved Bristle	*Othotrichum obtusifolium*
Moss, Bright Green Cave	*Cyclodictyon laetevirens*
Moss, Cordate Beard	*Barbula cordata*
Moss, Cornish Path	*Ditrichum cornubicum*
Moss, Derbyshire Feather	*Thamnobryum angustifolium*
Moss, Dune Threat	*Bryum mammillatum*
Moss, Glaucous Beard	*Barbula glauca*
Moss, Green Shield	*Buxbaumia viridis*
Moss, Hair Silk	*Plagiothecium piliferum*
Moss, Knothole	*Zygodon forsteri*
Moss, Large Yellow Feather	*Scorpidium turgescens*
Moss, Millimetre	*Micromitrium tenerum*
Moss, Multifruited River	*Cryphaea lamyana*
Moss, Nowell's Limestone	*Zygodon gracilis*
Moss, Rigid Apple	*Bartramia stricta*
Moss, Round-leaved Feather	*Rhyncostegium rotundifolium*
Moss, Schleicher's Thread	*Bryum schleicheri*
Moss, Triangular Pygmy	*Acaulon triquetrum*
Moss, Vaucher's Feather	*Hypnum vaucheri*
Mudwort, Welsh	*Limosella australis*
Naiad, Holly-leaved	*Najas marina*
Naiad, Slender	*Najas flexilis*
Orache, Stalked	*Halimione pedunculata*
Orchid, Early Spider	*Ophrys sphegodes*
Orchid, Fen	*Liparis loeselii*
Orchid, Ghost	*Epipogium aphyllum*
Orchid, Lapland Marsh	*Dactylorhiza lapponica*
Orchid, Late Spider	*Ophrys fuciflora*
Orchid, Lizard	*Himantoglossum hircinum*
Orchid, Military	*Orchis militaris*
Orchid, Monkey	*Orchis simia*
Pannaria, Caledonia	*Pannaria ignobilis*
Parmelia, New Forest	*Parmelia minarum*
Parmentaria, Oil Stain	*Parmentaria chilensis*
Pear, Plymouth	*Pyrus cordata*

Penny-cress, Perfoliate	*Thlaspi perfoliatum*
Pennyroyal	*Mentha pulegium*
Pertusaria, Alpine Moss	*Pertusaria bryontha*
Physcia, Southern Grey	*Physcia tribacioides*
Pine, Ground	*Ajuga chamaepitys*
Pink, Cheddar	*Dianthus gratianopolitanus*
Pink, Childling	*Petroraghia nanteuilii*
Pigmyweed	*Crassula aquatica*
Plantain, Floating Water	*Luronium natans*
Pseudocyphellaria, Ragged	*Pseudocyphellaria lacerata*
Psora, Rusty Alpine	*Psora rubiformis*
Ragwort, Fen	*Senecio paludosus*
Ramping-fumitory, Martin's	*Fumaria martinii*
Rampion, Spiked	*Phyteuma spicatum*
Restharrow, Small	*Ononis reclinata*
Rock-cress, Alpina	*Arabis alpina*
Rock-cress, Bristol	*Arabis stricta*
Rustworth, Western	*Marsupella profunda*
Sandwort, Norwegian	*Arenaria norvegica*
Sandwort, Teesdale	*Minuartia stricta*
Saxifrage, Drooping	*Saxifraga cernua*
Saxifrage, Marsh	*Saxifraga hirulus*
Saxifrage, Tufted	*Saxifraga caespitosa*
Solenopsora, Serpentine	*Solenopsora liparina*
Solomon's-seal, Whorled	*Polygonatum verticillatum*
Sow-thistle, Alpine	*Cicerbita alpina*
Spearwort, Adder's-tongue	*Ranunculus ophioglossifolius*
Speedwell, Fingered	*Veronica triphyllos*
Speedwell, Spiked	*Veronica spicata*
Starfruit	*Damasonium alisma*
Star-of-Bethlehem, Early	*Gagea bohemica*
Stonewort, Bearded	*Chara canescens*
Stonewort, Foxtail	*Lamprothamnium Papulosum*
Strapwort	*Corrigiola litoralis*
Turpswort	*Geocalyx graveolens*

Violet, Fen	*Viola persicifolia*
Viper's-grass	*Scorzonera humilis*
Water-plantain, Ribbon leaved	*Alisma gramineum*
Wood-sedge, Starved	*Carex depauperata*
Woodsia, Alpine	*Woodsia alpina*
Woodsia, Oblong	*Woodsia ilvensis*
Wormwood, Field	*Artemisia campestris*
Woundwort, Downy	*Stachys germanica*
Woundwort, Limestone	*Stachys alpina*
Yellow-rattle, Greater	*Rhinanthus serotinus*

ACKNOWLEDGMENTS

I am most grateful to Dr Michael Reiss at the University of Cambridge, Department of Education, and to Dr Roger Lock at the School of Education, Birmingham University for reading and commenting on this manuscript. I am also grateful to the Department for Education and the Department of the Environment generally. In particular the material used from the DES leaflet referenced below is produced by kind permission of the Controller of Her Majesty's Stationery Office.

REFERENCES

DES (1990) *Administrative Memorandum No 3/90*. London: Department for Education.

Lock R. (1989) Investigations with animals and the Animals (Scientific Procedures) Act 1986. *School Science Review*, **71**(255).

O'Donoghue P.N. (1988) The law and animals in schools. *Journal of Biological Education*, **22**(1).

Halsbury's Statutes (Fourth Edition, Volume 2, 1992 Re-issue; volume 32 and Cumulative Supplement 1993). London: Butterworth.

Halsbury's Statutory Instruments (Volume 2, 1993 issue). London: Butterwoth.

Halsbury's Laws (Fourth Edition, Volume 2 and Cumulative Supplement 1993). London: Butterworth.

Chapter 7   SAFETY

M. Ingram

ABSTRACT

Unlike the associated risks of using chemicals, the presence and handling of plants and animals should not pose a great threat to the teacher and pupil in school. However, there are recommendations that should be followed and these are discussed here. Safety policies exist for all Local Education Authorities, but individual schools in the maintained and independent sectors should develop their own policies to suit their conditions. The information regarding safety should be present in all schemes of work (see "risk assessments") so that any teacher will have easy access to the information. Details of these risk assessments are included, together with the details of the safe keeping of living organisms in schools. The main areas covered include aquaria, with an emphasis on electrical safety, greenhouses, where electricity is also important, the safe use of chemicals and glass, animals which can be a risk to asthma and bronchitis sufferers as well as posing a risk of *Salmonella* and tetanus, plants whose seeds or other structures can be poisonous and microbiology, with its inherent risks.

Although covering most of the school situations that are likely to arise, there are references to more detailed safety guidance which can be obtained from CLEAPSS, SSERC (the Scottish Schools Equipment Research Centre), the ASE (Association for Science Education), and DFEE (ex-DFE, ex-DES). In addition, there may be local regulations that apply to specific Local Education Authorities, and these take precedence over guidance from other sources.

SAFETY WITH ANIMALS AND PLANTS IN SCHOOLS

There are many excellent reasons for encouraging the use of living things in the classroom. However, the teacher needs to be particularly careful in the choice of suitable material, not only in its ease of culture or maintenance, but also from the perspective of the over-enthusiastic and inquisitive child - or teacher!

The extra work involved in caring for animals and plants, and the expertise that is sometimes needed, may be responsible for a reduction in the use of living material in schools, particularly secondary schools. With the increased emphasis on content-laden syllabuses, there are increasing pressures on teachers' time, with the further development that children may not be getting out of the classroom to visit natural ecosystems and plant/animal collections. This has effectively reduced pupils' contact with real, living things. There are also many more non-specialists involved in the teaching of biological topics nowadays, and they are recognising that it is far easier to set up a lesson with some batteries, wire and bulbs, then to provide some jars of beans showing two weeks of growth or three stages of locusthoppers at the drop of a hat!

With these points in mind, teachers need ready access to information, especially as they may be unfamiliar with the actual teaching concept, not to mention the technicalities of often working two or three weeks ahead to provide the necessary living stages. Legislation also changes, and what used to be possible is now not recommended, or in some cases usually forbidden - for example in the case of blood sampling. In the preceding chapters, reference has been made to specific aspects of safety. The purpose of this chapter is to provide a useful summary, additional, detailed guidance and sources of further reading if needed.

RISK ASSESSMENTS

When preparing science schemes of work for all teachers involved, in addition to the lesson content, etc., clear guidance must be given on any risk to the teacher and/or pupils. The assessment of the risks involved in any procedures in science teaching is a requirement under the Management of Health and Safety at Work and Control of Substances Hazardous to Health (COSHH) Regulations, and it is the responsibility of employers to give instructions on how this should be achieved. Most have adopted the principle of requiring schools to consult *general risk assessments* to determine what is permitted and the precautions etc. that are needed. These general risk assessments include materials from CLEAPSS (see below), the ASE (*e.g. Be Safe!* (for primary schools) and *Topics in Safety*), SSERC and the DFE (*e.g. Microbiology: an HMI guide for schools and further education*). This is often limited to safety when using chemical or sharp instruments, etc. It would be quite possible to omit less obvious, and yet just as important, risks such as diseases transmitted by tortoises; or the toxic nature of seeds. Safety recommendations are constantly being changed, so teachers should make efforts to re-appraise themselves of the latest guidance.

Risk assessment is not just about identifying the problem; it is more about what to do to avoid the risk in the first place or, if a problem arises, what should be done about it. CLEAPSS documents outline these very well indeed and form the basis of any risk assessment that may be needed in most school situations.

RISK ASSESSMENT PROCEDURES

In carrying out their planned operations, teaching staff must compare these with what is in the general risk assessments of safety guidance texts and modify their procedure accordingly. The role of employers in providing employees with risk assessments has already been referred to, and all authorities and individual schools or colleges will have, or should be in the process of producing, safety policies for their staff and students.

Schools must assess risks in the following way, unless the employer has given alternative instructions.

1. Initially search out and consult relevant general risk assessment texts.
2. Safety warnings should be written into published materials used by teachers as these do not always have the necessary guidance.
3. Science departments and science teachers should write safety information into "point of use" materials. These will include work-schemes, technicians' notes, lesson plans, as well as work sheets and textbooks if practical.

There will obviously be different approaches to suit individual circumstances, but whatever the process, there must be evidence that staff have given the process some thought and have not just been asked to refer to the safety texts which are notoriously lengthy tomes.

Some useful checkpoints could be kept at hand to fill out as schemes of work or work sheets are processed:

Check ... Is it advisable to do this experiment or activity with this age group?
Check ... If the substances to be used are toxic or corrosive, can alternatives be used or changes made to the procedure?
Check ... Can the operations be made safer for the students?

Whatever warnings are made in written texts, the teachers must give oral warnings too and supervise all pupils carefully.

An example of a work sheet produced for students to follow when carrying out a set of experiments in microbiology is given in Figure 7·1.

Figure 7·1. An example of a worksheet

MICROBIOLOGY - IMPORTANT

Our first set of experiments are designed to illustrate some of the simple principles of growing bacteria but before carrying them out there are a few rules which must be understood and learned.

Laboratory Hygiene and Aseptic Technique

1. Although many microbes are harmless you do not know which ones are not so treat them all as though they are dangerous.

2. Do not eat or put your hands to your mouth whilst in the laboratory.

3. Always wash your hands thoroughly before leaving the laboratory and use the disposable paper towels for drying.

4. Pots of strong disinfectant which is caustic and can be irritating and painful to the skin are provided for disposing of certain material which is finished with. You will be told when to use them for disposal.

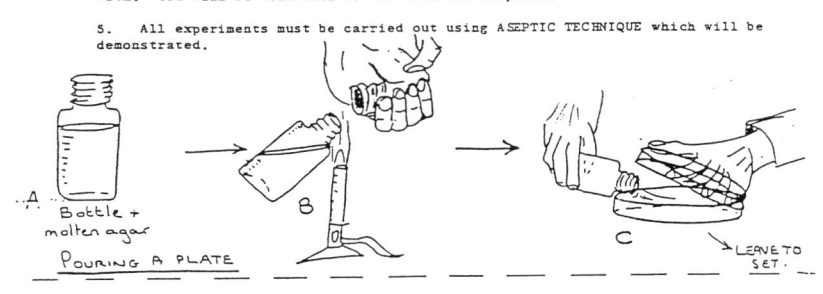

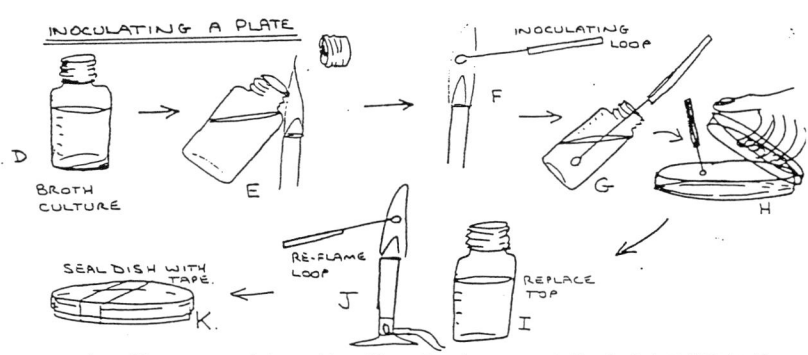

5. All experiments must be carried out using ASEPTIC TECHNIQUE which will be demonstrated.

A Bottle + molten agar
POURING A PLATE
B
C LEAVE TO SET

INOCULATING A PLATE
D BROTH CULTURE
E
F INOCULATING LOOP
G
H
SEAL DISH WITH TAPE
K
J RE-FLAME LOOP
I REPLACE TOP

6. If any material is accidentally spilt please report the fact immediately to your teacher who will deal with it.

The technician's list or lesson notes for preparation of equipment, etc. could look something like this:

Year 10. Topic Microbes

Lesson 5. Title: Where are microbes found?

Each class will need:

15 Petri dishes, containing nutrient agar

OHT pens for writing on dishes

Sterile cotton buds, in aluminium foil

Sellotape dispensers x4

Pots of disinfectant x4 (for each bench)

JeyCloths for spills x2

Inoculating loops x15

Please sellotape a pile of dishes x 5 so that they cannot accidentally fall over and lids become knocked off as the groups arrive!

Also check with the teacher that they have read the Risk assessment regarding where to get samples from ... *NB* not the loos! or scabs on their fingers! Yuk.

At the end of the lesson they SHOULD have all sellotaped up the dishes. Could you check before they go into the incubator that each has 2 small strips of sellotape on the edge? If any have put it all round the rim, could you remove it and change to the advised method?

Incubator should be set at 30 °C and dishes incubated upside down.

Further note:

Once the agar plates have been incubated, please then seal each dish around the circumference with sellotape. This will help to stop inquisitive fingers detaching one of the initial strips of tape in order to have a good look! An example of the scheme of work which is to be used by the teacher, may look something like this:

> Lesson 5. Microbes.
>
> In this lesson, you will be introducing them to the idea that microbes are found everywhere. The idea is to show them the dishes of agar which you have already discussed. Explain to them that these are sterile (meaning ...?) and must not be opened until they know what to do. Ask for some ideas of where there might be microbes. RISK: you must make it clear that it would be very dangerous to collect samples from the toilets or from spots on their skin so they are NOT to do this.
>
> Some good ideas to try are:
>
> Dust from window sills, water from the fish tank, soil, just open to the air for 1 minute, a piece of hair, a coin touched on the surface, fingers before and after washing .. the latter is good since often there are more microbes after washing as they are dislodged from the creases.
>
> When they have decided, show them the sterile loop and cotton buds techniques again. Important: sellotape up each dish, with only two bits on the side *not* all the way round the circumference. Put in tray on front ...

Although risk assessments may seem a tedious exercise, if this is an ongoing process the work can be completed in a relatively short time, and could be the focus for an INSET session, when several staff can be involved.

When complete, the risk assessments must be reviewed if new restrictions are imposed or if work schemes change. It is always important to consider how a particular class of pupils might behave and this could lead to further restrictions or modifications of procedures.

RISKS IN SCHOOL BIOLOGY WHEN USING LIVING MATERIAL

In general, using living material in schools may present two main forms of hazard:
 a) the effect of the living organism or its products upon staff and pupils
 b) physical hazards, such as the equipment and chemicals used in handling such living things.

The information given here covers the majority of commonly encountered hazards, but teachers must consult local advice from their employers (LEA or governors) as there may be local rules and regulations which apply.

AQUARIA

There are three hazards connected with keeping aquaria:
 a) electrical fittings
 b) the glass
 c) the weight of water.

It should be obvious that electricity and water should be used with great care, and electrical fittings for aquaria must meet the highest standards of safety. Metal frames and hoods must be earthed. New laboratories in secondary schools will be fitted with RCCB's (residual current circuit breakers) (Figure 7·2).

Figure 7·2. Residual Current Circuit Breakers

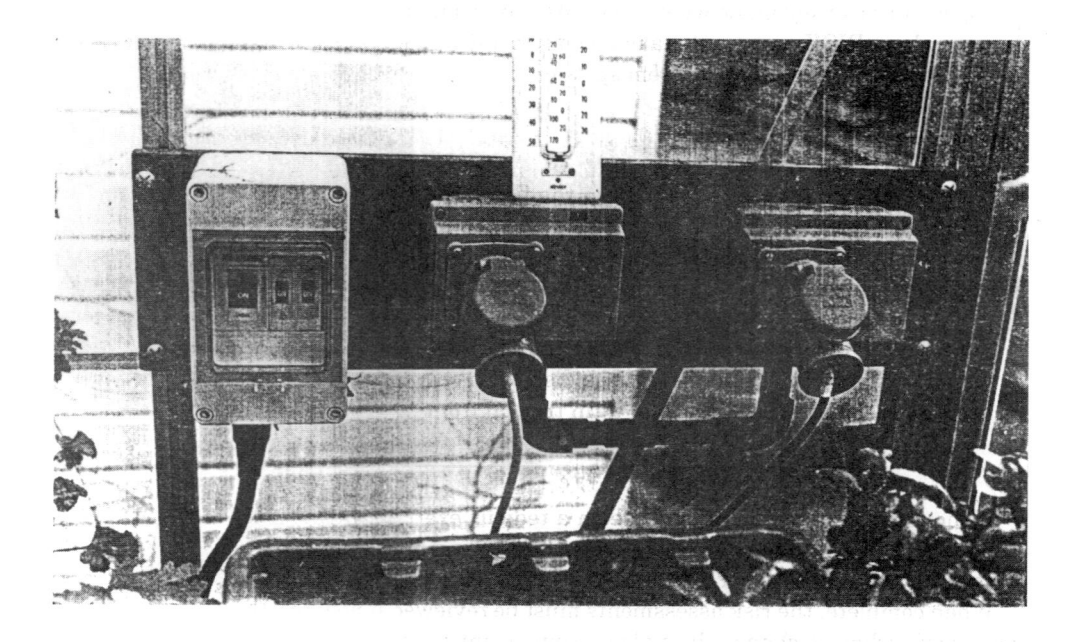

These will automatically switch off the current within a fraction of a second if accidental contact with live current is made, thus preventing electrical shock if a fault develops in the system. However, it is unlikely that classrooms in primary schools, or old laboratories, will have these devices. It is essential, therefore, to use a similar RCCB that can be obtained from most high street suppliers. These are usually sold for running off a normal 13 amp socket. All connections should be waterproof and the best lighting to use is a fluorescent tube, rather than filament bulbs which get hot, and allow water to enter their fittings more easily.

Many new aquaria are all-glass construction, and care must be taken with edges which should be rounded off or covered to prevent cuts. A full aquarium is very heavy and should not be carried or moved until emptied. When moving a glass aquarium, gloves should be worn, to avoid cuts should it slip.

MICROBIOLOGY AND BIOTECHNOLOGY

The use of microorganisms in school is an interesting and valuable teaching strategy, but these "invisible" organisms can present some of the greatest risks to staff and students. It is essential, therefore, that all staff are trained in their use and that students are similarly given very clear and correct instruction when handling microbes. It is unlikely that primary schools will have facilities for work with agar plates but simple activities with yeasts, yogurt, etc. are quite feasible (see chapter 2).

All purchased cultures must be from reputable suppliers.. Growing microbes for subculturing must only be done with "known" cultures. Finding out what microbes are present in, for example, the air or investigating the presence of microbes on fingertips is fine, but the results must not be investigated further by opening dishes and gaining access to the colonies of microbes that have been grown. Any counting or observation of colonies is easily and safely carried out through the lids of dishes, which should have been completely sealed around the circumference with tape, *after* incubation.

Under no circumstances should the source of study include potentially hazardous environments such as toilet seats or human fluids such as saliva. Coughing on culture plates and then incubating potentially hazardous microorganisms is an irresponsible practice and must *not* take place.

When incubating microbes, the temperature should be at ambient temperature, and not higher than 30 °C to discourage the growth of potentially harmful human pathogens. In many cases, simply incubating in a designated area of the preparation room is quite sufficient.

Avoid the use of special, selective culture media, such as blood agar. The use of nutrient and malt agar media is probably all that is needed for much microbiological work. Usually the instructions given in textbooks to accompany a certain practical will include these media. Older textbooks may advocate, for example, the use of MaConkey's agar, which is used for the growth of coliform bacteria typically found in sewage. These types of selective media should not normally be used. Starch and milk agar are, however, quite safe. Broth cultures are to be treated in the same way as solid, agar-based media.

Culture media	Organisms recommended	Context
Nutrient broth	setting up stock cultures; *Escherichia coli; Bacillus subtilis*	growing microbes for noculating plates
Nutrient agar	most microbes; including *Escherichia coli; Bacillus subtilis*	most demonstrations of growth of microbes
Malt agar	yeasts and moulds	selective for fungi
Milk (lactose) agar	yogurt/milk	milk freshness and demonstrating the presence of "helpful" microbes

When conducting practical work, particular attention should be drawn to hygiene. Eating should be prohibited and great care taken with disinfecting the work area, especially after completing the work. Despite their appeal, eating the products of such practical work (such as yogurt, cheese and wine) in the laboratory must be prohibited. If activities involving the making of bread and yogurt can be transferred to an area usually devoted to cooking, such as a food technology room, then eating the product may be acceptable. It will be necessary to keep the utensils required entirely separate from laboratory equipment and for them to be cleaned in a domestic dishwasher that is not used for laboratory equipment. Hygiene in a general purpose laboratory cannot be relied on.

DEALING WITH SPILLS AND DISPOSAL OF CULTURES

A number of proprietary disinfectants are available from educational suppliers. There is no single disinfectant suitable for all purposes. For work with bacteria and fungi, a clear phenolic disinfectant is recommended as this has a wide spectrum of activity and is not degraded by organic matter. (This type of disinfectant is, however, ineffective against viruses.) Hypochlorite disinfectant (minimum strength 1% chlorine) is effective against all microorganisms, including viruses, but is easily broken down in dirty conditions or in contact with organic matter. Biocidal Ampholytic Surfactant-type disinfectants, such as ASAB, BAS or Tego are satisfactory for some tasks such as wiping down bench surfaces but are not now thought to be as widely effective as the disinfectants discussed above. It should be noted that a minimum contact time of 15 minutes is required for adequate disinfection, unless ethanol is used, which is effective within 5 minutes.

Care must be taken with all disinfectants, especially the concentrates. Gloves and eye protection must be worn when decanting the concentrates, and when wiping surfaces. In all microbe work, pots of disinfectant should be available for the disposal of pipettes etc and to deal with spills which should be tackled immediately.

Plastic Petri dishes and other disposable apparatus should be placed in roasting bags (or autoclavable plastic bags) and then steam sterilised in a pressure cooker used solely for the purpose, or an autoclave. These pieces of equipment should be checked every 12 months by a competent person according to the procedure organised by the employer.

When using pressure cookers or autoclaves, care must be taken to avoid scalding. The pressure must be allowed to reduce gradually, and the valves not lifted prematurely. Articles in a recently-opened pressure cooker will be very hot and thermal gloves may need to be worn or time given for the contents to cool down to a safe temperature.

After sterilising, agar plates and other plastic articles can be disposed of safely in dustbins or incinerated. Some schools may wish to liaise with their local hospital which may be able to help them with disposal if there are large quantities. The problem with this is the inevitable build-up of sacks waiting for collection and this must be carefully considered.

LABORATORY PROCEDURES

A common error in culturing bacteria is for students to be asked to sellotape the dishes completely around the rim once they have been inoculated. Although this deters students from opening the dishes, the exclusion of air could encourage the growth of anaerobic organisms, some of which are very hazardous to health. It is much better to place two strips of sellotape over the sides, allowing normal ventilation of the lid and for the plates to be sealed completely, where appropriate, **after** incubation before they are examined by pupils.

BLOOD AND CHEEK CELL SAMPLING

The DFE advises against blood cell sampling on students and this advice is to be heeded for work with most pupils. Most LEAs have prohibited taking of blood samples. This advice also extends to cheek cell samples, but the Institute of Biology recommends a procedure to be followed which *is* safe:

> *"Using a fresh cotton bud from a newly opened pack, the cheek is rubbed gently on the inside. It can then be smeared over a small area of a clean microscope slide, after which the cotton bud is placed into a hypochlorite disinfectant, such as hypochloride or ethanol. After drying, the slide can then be stained with aqueous methylene blue, covered with a cover-slip and examined using a microscope. Used slides are then disposed of in hypochlorite."*

Several LEAs permit the use of this procedure but many do not and until their rules have been amended even the procedure given above cannot be used. Alternatives have been recommended, including scraping the inside of a sheep's trachea, or removing cells from human skin using sellotape, but both these produce a preparation which is inferior in quality and some students may find the handling of "dead animal bits" to be unacceptable.

SALIVA

Due to obvious health risks, the use of saliva for enzyme work should follow scrupulous hygiene procedures. It should be noted that the DFEE has **not** advised against the use of saliva. Alternatively, the use of amylase, which can be obtained from biological suppliers, can be considered. Proprietary amylase is also more predictable than the variable strength of saliva from different individuals, which is dependant on several factors such as recent meals. This may be an advantage when doing practical experiments, but a disadvantage when conducting more open-ended investigations. One solution is for students only to use their own saliva in any work and to wash up their own glassware. Commercial amylase usually, however, contains reducing sugar (so invalidating many control experiments) and there have been reports of staff developing allergies to this enzyme when used in large quantities.

GREENHOUSES

As with aquaria, the obvious hazard in greenhouses is broken glass. In addition, heating, electrical systems and chemicals must be considered as hazardous. Greenhouses should be constructed to a professional standard, and sited where risk is reduced. As thin, horticultural glass is often used, it may be worth considering less fragile plastic sheets as an alternative, especially at ground level if children pass nearby. Broken glass must be replaced immediately, and disposed of sensibly to avoid accidents.

Electrical systems

As with all electrical systems which are outside or used in the with proximity of water, a RCCB (residual current circuit breaker) must be incorporated - preferably as a permanent part of the distribution system for the greenhouse. Any heating or lighting should be installed professionally, using waterproof sockets and switches. Portable heaters should be firmly fixed, to prevent accidental movement. Open element heaters must **not** be used. Waterproof sockets (electrical) should be used for supplementary lighting or propagators within the greenhouse.

Chemicals

All chemicals used on plants must be used with care, and instructions followed explicitly. This applies not only to pesticides (such as insecticides and fungicides) but also to fertilisers in dry form.

It would be prudent to have a basic first-aid kit handy for splashes in the eyes, and cuts. The provision of water in the greenhouse not only makes watering easier, but provides the opportunity to wash off splashes of chemicals. Eating in greenhouses and chemical storage areas should be discouraged, as hand to mouth contact could transfer chemicals.

All pesticides and herbicides should be locked in a poisons cupboard. Some fertilisers are oxidising agents, and should be stored away from other chemicals, preferably in fire-proof containers. Avoid keeping large quantities at any one time.

ANIMALS

There are few risks to health from invertebrates, though locusts kept permanently in classroom laboratories or preparation rooms can pose a threat of allergies developing due to the dust they create. Teachers may be at a greater risk if in longer contact with such animals. Wearing dust masks and protective gloves is advisable when cleaning out. Certain invertebrates available for class studies are easy to breed and, as a result, are often pest species in granaries and food storage areas. Floor beetles (used in genetics), cockroaches and blowfly maggots must not be allowed to escape where they could become a health risk.

Recently giant African land snails have become widely available in pet shops. There is a risk of imported ones carrying a parasitic lungworm which can infect humans. Any snails bred in the UK (the vast majority) will, however not be parasitised. Obtaining these snails from reputable sources is obviously important.

Keeping bees in school is a worthwhile and highly educational exercise. However, great care and attention is needed as this is not for the inexperienced. The siting of any hives must be carefully considered to avoid children and adults straying into their flight paths. Staff, mowing and tending to school grounds, can be particularly at risk as their noise and activity can disturb bees and may cause them to sting. Bee stings can be dealt with using antihistamine preparations (although excessive use of antihistamines must be avoided). Severe reaction to a bee or wasp sting can be serious for some individuals and medical help must be sought immediately in such cases. If children are to experience visits from local beekeepers (who are always keen to bring a demonstration hive or a nucleus which can be used to set up a new colony) then a letter home to parents for advice on any known allergies may be prudent.

Vertebrates

The usual animals kept in schools offer little risk, but some more exotic species are available from local pet shops and these may cause problems - such as terrapins and tortoises which may be infected with the *Salmonella* although they themselves show no symptoms. It should be noted, however, that reptiles and several other animals such as amphibians may also carry this organism. Stocks obtained from reputable sources are less likely to be infected, but good hygiene, which should always be observed, will normally overcome problems of disease transmission.

Cage birds, such as members of the parrot family, including budgerigars, can carry psittacosis, other species may be carrying the similar disease ornithosis. Care must be taken to practise stringent hygiene. The DFEE recommends that these should *not* be kept in schools.

Some reptiles, especially exotic snakes, are difficult to keep and the risk from venomous snakes is obvious!

Occasionally, pupils or staff become sensitised (or *are* already sensitive) to fur, feathers or droppings. Symptoms include skin rashes, dermatitis or asthma. Any small mammals may produce such allergic reactions, and this is especially so with those whose daily routine brings them into contact with dust and fur from these mammals. Teachers should be alert to the development of such allergies and when they become apparent, individuals are advised to avoid contact with the mammals.

It goes without saying that all animals should be kept in adequate housing which prevents invasion by wild animals such as mice, rats and foxes. Zoonoses (diseases transmittable from animals to humans) are a risk in such situations. The keeping of native British wild birds and mammals is obviously to be avoided so care must be taken if dead or injured animals are brought in by children. Such animals as these must be kept isolated, handled with suitable protection and taken to a local vet, RSPCA or PDSA treatment centre as soon as possible.

Before and after handling animals, all staff and pupils should wash their hands. Allowing students to take animals home is acceptable in certain circumstances (*e.g.* school holidays) but permission must be granted from parents and efforts made to avoid contact with wild species. For example, keeping guinea pigs in an outside enclosure at home, in summer, may expose them to such wild contacts and so should be avoided. Occasionally children may bring in their pets from home - care is needed to avoid contact with school stock. Dogs and cats must be healthy and up to date with their particular immunisation programme.

Tetanus

Anyone handling animals regularly or involved in regular contact with soil outside (e.g. gardening clubs) should be immunised against tetanus. A letter to parents advising them to seek protection for their children is recommended.

PLANTS

It is surprising how many plants' parts can, in theory, be a problem. Daffodil bulbs, if eaten, can be toxic. In reality, the risks are probably not great as the taste of such material if often unpleasant! However, quite innocent looking berries and seeds can be confused with fruit harvested in the garden and so eating fruits or seeds as part of any science work should be prohibited.

The main risks are through poisoning by eating certain plant parts and through ingesting chemicals used on seeds as fungicides or pesticides.

THE USE OF METHANAL/FORMALDEHYDE IN PRESERVATION

While the emphasis in this book is on living animals and plants in schools, it is recognised that preservation of specimens may be seen as a useful teaching aid. One major concern to safety is the use of methanal (formaldehyde) as such a preservative. This toxic substance is purchased at 40% strength - at which concentration the vapour and liquid is extremely dangerous to eyes. Permanent damage may occur in the time taken to reach a tap to wash the eye out. If methanal *is* used, eye protection should be worn at *any* concentration. This also applies to material that is preserved in methanal for dissection purposes. However, alternative safer preservatives are now available and the use of freshly-killed or recently-frozen specimens is by far the best way to keep material for dissection activities. (*N.B.* Such material must, however, *not* be kept in a fridge or freezer used for substances consumed by humans.) Note, however, that there are no substitutes for the use of methanal for fixing tissues. Certain gardening books still advocate using "formaldehyde" for washing down glass and staging. Bearing the above risk in mind, this is inadvisable, and household disinfectant seems a better alternative - especially in an area where overhead cleaning is likely to take place.

Treatment of splashes

Should splashes of methanal/formaldehyde enter the eye, treat by washing with a continuous stream of cold water for at least ten minutes, after which medical attention should be sought. If any is swallowed, the casualty should be encouraged to drink plenty of water, and medical attention should be sought.

EXPERIMENTS USING STUDENTS AS SUBJECTS

Under no circumstances should children be forced to take part in experiments which use themselves unless they want to. They should not be made to feel inferior, and the spirit of competition should be avoided as this could lead to unacceptable risks or stress. The main problems encountered are with asthma and bronchitis sufferers. The DFE has suggested that, as it is difficult to identify students at risk, it is inadvisable to carry out lung ventilation experiments but this approach would prevent all simple investigations on breathing rates in which a little exercise is taken. Seeking parental permission may be a suitable approach to overcome this, if the experiment is carefully supervised at all times.

A minor risk to students is in studies of variation involving tasting of PTC (Phenylthiocarbamide). No attempt should be made to taste more than two paper strips impregnated with the chemical, as the accumulated effect can be toxic. Mapping areas of the tongue for taste sensations should be carried out using sugar, citric acid, salt and cold coffee or tea, rather than the quinine recommended in some older textbooks. Such taste testing investigations must only be permitted if special arrangements are taken to guarantee hygienic conditions.

Data logging and monitoring of students' reactions or pulse rates should only be made with proprietary equipment, using a battery-powered supply to the electrodes. No attempt should be made to adapt these to run off mains electricity.

CONCLUSION

Reference to *safety* in aspects of science can tend to discourage teachers from organising worthwhile, valuable areas of learning experience. Although this chapter tends to emphasise the "negative" aspects, and could be frightening to the uninitiated, it is worth stressing that many of the procedures referred to here are easily incorporated into standard procedures and should not be seen as over restrictive in any way.

RESOURCES AND FURTHER READING

ASE (1987) Safety and the School Pond. *School Science Review*, **69**(247). pp 286 - 289
ASE (1988) *Safeguards in the School Laboratory*. Hatfield, Herts: Association for Science Education.
ASE (1988) *Topics in Safety*. Hatfield, Herts: Association for Science Education.
ASE (1990) *Be Safe!* Hatfield, Herts: Association for Science Education.
CLEAPSS (1992) *Laboratory Handbook*. London: Brunel university (available, with Guides and Hazcards, only to members of CLEAPSS).
Cooper M. R. and Johnson A. W. (1988) *Poisonous Plants and Fungi: An Illustrated Guide*. London: Ministry of Agriculture, Fisheries and Food.
COSHH (1989) *Guidance for Schools*. London: HMSO.
McQuillan P. (ed.) (1988) *Croner's Manual for Heads of Science*. London: Croner Publications Ltd.
DES (1990) *Administrative Memorandum No 3/90*. London: Department for Education.
DES (1990) *Microbiology: an HMI Guide for School and Non-Advanced Further Education*. London: HMSO.
DES (1985) *Safety in Science Laboratories - DES Safety Series No 2*. London: HMSO.

USEFUL ADDRESSES

RSPCA,
The Causeway,
Horsham,
West Sussex,
RG12 1HG.

UFAW (Universities Federation for Animal Welfare),
8 Hamilton Close,
South Mimms,
Potters Bar,
Herts,
EN6 3QD.

Chapter 8   MORAL AND ETHICAL ISSUES

R. Lock and M. Reiss

ABSTRACT

Pupils often hold strong views about how society should treat animals. However, mismatches frequently exist between pupil attitudes, pupil behaviour and pupil knowledge. Some people argue that the use of animals for human purposes is wrong in itself. Others argue that whether or not the use of animals for education, or any other purpose, is acceptable depends on the consequences of that use. Whether animals have rights is a contentious matter. However, the central point at issue can be discussed without it being necessary for someone to declare themselves either for or against the notion of animals' rights. We conclude that animals can suffer and that teachers have a duty to minimise any such suffering by animals used for educational purposes. At the same time, though, we feel that biology teachers have a duty to include the natural environment and living organisms, including animals, in their teaching. An attempt is made to suggest how teachers might begin to balance the advantages to their pupils or students of using animals in their teaching against the possible disadvantages to the animals of such use. Specific recommendations are given about dissection in schools and about the teaching of controversial issues, including the issue of how people should use animals.

INTRODUCTION

This chapter looks at the ethical implications of using animals in education. It begins by examining what pupils in secondary schooling feel about the rights and wrongs of society's use of animals in research, education and farming. It then goes on to construct an ethical framework within which our use of animals can be considered. The arguments for and against dissection are considered and recommendations are made about dissection in secondary schools. Finally, we conclude by suggesting how controversial issues, such as the question of our use of animals, can be taught in science. Perhaps inevitably the focus is on 11 to 18 year olds because it is with this age range that the ethical issues are most acute and it is on this age group that research has been done. Nevertheless, we hope that this chapter will be of value to primary teachers.

PUPIL PERCEPTIONS AND ATTITUDES: WHAT ARE PUPILS' ATTITUDES TO THE USE OF ANIMALS?

Pupils come to their science lessons with all sorts of existing views about how animals, plants and the physical environment should be treated by humans. Considerable diversity of pupil opinion exists, as we shall discuss below. Nevertheless, many pupils are deeply suspicious about the ways in which adults treat the natural world. In this section we will focus on pupils' attitudes to the use of animals, as these raise the most challenging and immediate moral issues for biology teachers.

Chapter 8   Moral & Ethical Issues

An extensive survey in this country of pupils' attitudes to the use of animals in research, education and farming was recently carried out on a sample of 468 Year 10 pupils of mixed ability (Lock and Millett, 1991). These pupils came from 10 state secondary schools in England and completed a detailed questionnaire in science/biology lessons. A selection of the findings from this study is presented in Table 8·1.

Table 8·1

14-15 year olds' attitudes towards animal experimentation, animals in school, and the use of animals in farming; dotted lines separate statements relating to these three categories (percentage of females and males responding to each category). (Taken from Millet and Lock, 1992.)

Attitude statement	Agree		Uncertain		Disagree	
	F	M	F	M	F	M
New medicines should be tested on animals before they are taken by humans.	12	25	26	24	62	50
I believe in a total ban on animal experiments.	60	41	23	32	17	28
I would take a medicine that had been tested on animals if it would save my life.	63	85	31	14	7	4
Research from animal experiments improves the lives of people.	17	38	43	41	40	21
I think that no more cosmetics product like shampoo or lipstick) should be tested on animals.	88	71	6	14	6	15
A new washing up liquid should be tested on animals' skin before being sold in the shops.	4	10	9	18	87	73
Medicines used for treating pet dogs should be tested on laboratory dogs first.	18	35	48	39	3	25
I would find dissecting (cutting apart) a dead animal interesting.	18	45	16	24	66	31
I don't think that woodlice come to any harm if they are used in investigations to find out what conditions they prefer.	48	59	35	29	16	12
I would dissect a part of an animal that had been killed for food (e.g. sheep's eye).	26	57	27	19	47	24
It is a good idea to keep animals in school to watch how they live.	36	49	31	25	32	26
I think that watching earthworms to see how fast they crawl is cruel.	31	15	32	34	36	61
I would rather learn about animals from books than by watching live animals.	31	27	32	18	35	55
I would not mind watching garden snails moving to find out more about them.	72	74	21	17	8	9
I believe it is wrong to kill animals for food.	35	12	32	23	33	65
Keeping animals to provide food, (eggs, milk) for people is acceptable to me.	77	88	16	8	6	3
Farmers should keep sheep for their wool and not for their meat.	60	34	24	30	16	36

A number of conclusions, some of them illustrated by the data in Table 1, can be drawn from this research which extends earlier work by Furnham and Pinder (1990) and others:

- Pupil attitudes about the use of animals vary from strong agreement to strong disagreement, depending on the particular use of the animals in question.
- Males are more likely than females to condone the use of animals for human benefits.
- In general, the following trend is discernible with regard to the approval by pupils of animals in particular contexts (Lock, 1993):

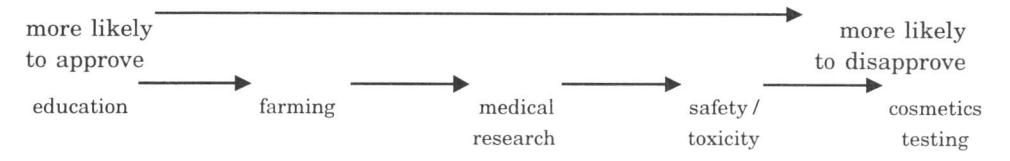

more likely to approve					more likely to disapprove
education	farming	medical research	safety / toxicity	cosmetics testing	

- There is often a substantial mismatch between pupil attitude and behaviour. For instance, 35% of the girls and 12% of the boys agreed with the statement "I believe it is wrong to kill animals for food", yet 87% of the girls and 98% of the boys said that they ate chicken, or mammals and chicken, or mammals, chicken and fish.

- There is often substantial pupil ignorance about why animals have been or are used in research. For instance, when asked "What medical advances have come from research using animals?", the answers in Table 8·2 were recorded. It is evident that a large majority of pupils had no idea about how research on animals has contributed to medical advances.

Table 8· 2
14-15 year olds' answers to the question: "What medical advances have come from research using animals? (Taken from Lock and Millett, 1992.)

Medical advances in treating or developing	Number of respondents	
	Females	Males
Cancer	19	20
Polio vaccine	5	4
Smallpox	1	4
AIDS	2	3
Diabetes	2	
Transplants	1	
Rubella	0	
Don't know	159	
No response	9	
Uncodeable		

Nothing seems to have been published about the attitudes of primary school children to the use of animals. Stanisstreet, Spofforth and Williams (1993) compared the attitudes of 147 11-12 year olds (53% female) and 137 15-16 year olds (45% female) in a recent study of three secondary schools. In all, significant differences between the two age groups were found in responses to 9 of the 27 statements included on a questionnaire (Table 8·3). As a generalisation, the younger pupils felt more strongly about the issues.

Table 8·3
Comparison of the responses of 11-12 year old and 15-16 year old pupils to statements concerning the use of animals. (Taken from Stanisstreet, Spofforth and Williams, 1992.)

Abbreviated statement	Pupils giving positive responses ('strongly agree' or 'agree') per cent	
	11-12 year olds	15-16 year olds
Wrong to wear leather shoes and jackets	30	12
Wrong to kill animals for their skins	80	66
Wrong to keep animals as pets	17	5
Wrong to experiment on animals for medical research	61	45
Wrong to dissect dead animals for teaching	51	37
Wrong to test cosmetics on animals	85	75
Wrong to kill fish for food	34	16
Wrong to kill animals for food	50	24
Wrong to keep chickens in battery cages	79	65

Before going on to examine the implications of such findings for biology/science teachers, it is appropriate to attempt to set the concerns shown by pupils, and adults, about the use of animals in an ethical framework.

AN ETHICAL FRAMEWORK FOR THE USE OF ANIMALS

One of the unfortunate features of the debate about how humans should use animals has been the polarisation of arguments. Differing views are all too often stereotyped and exaggerated. The result is that people who may hold not altogether dissimilar positions with regard to the well-being of non-human animals (henceforth referred to as 'animals') are portrayed, and may even see themselves, as irreconcilably opposed.

Before beginning to outline an ethical framework for the use of animals for human benefit, an important moral distinction needs to be made. Some argue that the use of animals for any purpose is wrong *in itself*. This is an *intrinsic* argument. Others take the view that whether or not the use of animals is acceptable depends on the *consequences* of that use. This is an *extrinsic* argument. The distinction is important. Someone who believes that the use of animals for human ends is intrinsically wrong will not be swayed ~~m~~ments about the value of their use to humans. As the scientific and educational ~~erally attach more weight to extrinsic than to intrinsic arguments, it is ~~ificance of the latter is appreciated. If not, there is little hope for ~~ with different views on this issue.

ANIMAL RIGHTS AND SPECIESISM - THE INTRINSIC ARGUMENTS

Whether animals have rights is a contentious issue (Regan, 1984; Tudge, 1988; Smith and Boyd, 1991; Carruthers, 1992; Dunayer, 1992; Garner, in press). However, the central point at issue can be discussed without it being necessary for someone to declare themselves either for or against the notion that animals have rights. The argument that the use of animals for human ends is speciesist (analogous to racism and sexism) proceeds as follows (*e.g.* Singer, 1975; Midgley, 1983): for society to permit animals to be used without allowing humans similarly to be used is morally indefensible. The fact that humans belong to a different biological species from, say, chimpanzees, dogs and laboratory rodents, is irrelevant.

What is at issue is whether animals can suffer (as first pointed out by Bentham, 1789). Suffering involves susceptibility to pain and an awareness of being, having been or being about to be, in pain. It is difficult to argue against the contention that vertebrates, and possibly certain invertebrates, can experience pain (Bateson, 1991, 1992). The extent to which animals are aware of their pain is more open to question. There is little doubt that certain of our closest evolutionary relatives, specifically the chimpanzee, have the requisite degree of self-consciousness (Dawkins, 1980; Humphrey, 1986). Although the extent to which other animals can suffer is still contentious, a growing number of biologists and philosophers accept that, at the very least, most mammals, and probably most vertebrates, can. Mammals are, of course, widely used in education, farming and research.

As we are still considering only intrinsic arguments against the use of animals, let us assume that some animals, at least, can suffer. The argument from someone who refuses to accept the use of such animals if it harms them, whatever the benefit to humans or other animals, is close to completion. (S)he argues that we only allow humans to be 'used' in, for example, research, provided two conditions are met. First, that the individual consents; secondly, that there is no intent to do harm. The application of these criteria to animals would, of course, cause all such research to cease. Further, it can be argued that if one adopts solely the criterion of suffering to decide whether or not an organism should be used in research, not only would the use of most laboratory animals cease, but a case could be made for mentally-handicapped newborn human infants to be used for such research, on the grounds that such infants are arguably incapable of suffering yet physiologically closer to self-conscious sentient people than the laboratory animals presently used.

In the 1991 Report of the Working Party of the Institute of Medical Ethics into the ethics of using animals in biomedical research, it is argued that this, to most people, unacceptable conclusion can be rejected on the grounds that:

> 'being of the human species may be a *sufficient* condition of being awarded enhanced moral status ... Possessing the *nature of a rational self-conscious creature* may be sufficient for being awarded this status even though his nature be impaired or underdeveloped in the individual case.'
>
> (Smith and Boyd, 1991, p 323, original italics)

To someone who believes in animal rights such an argument is special pleading. Consider a person who believes that society's use of animals in research and education is speciesist. It is precisely the argument of such a person that belonging to the human species is, on its own, irrelevant (Reiss, 1993a).

COST/BENEFIT ANALYSIS TO HELP DECIDE WHETHER ANIMALS SHOULD BE USED FOR A PARTICULAR END - THE EXTRINSIC ARGUMENTS

It is time to discuss the extrinsic arguments about the use of animals for human ends. Do the benefits (positive consequences to humans) outweigh the costs (negative consequences to animals)? This way of looking proceeds on a case-by-case approach. Patrick Bateson (1986) suggests that a piece of medical research involving the use of animals could be ranked on three axes: a 'Quality of research' axis, a 'Certainty of medical benefit' axis and an 'Animal suffering' axis. Each axis runs from low to high and the position of a point inside the cube defined by these three axes determines whether or not the use of animals should be permitted in the research.

For instance, a high quality piece of medical research with a high certainty of medical benefit might be permitted even if the animal suffering caused was high. On the other hand, Bateson argues, a piece of research of average quality, and with only a low certainty of medical benefit, should not be permitted even if the animal suffering caused was low (Bateson, 1986). Bateson has refined his model subsequently (Smith and Boyd, 1991) but the fundamental principle of weighing costs and benefits remains.

The most obvious problem with this approach is not so much in deciding where a piece of research lies on these three axes, but on deciding how to balance the costs (animal suffering) against the benefits (quality of research and medical benefit). However, a similar procedure has now been used for some years by the Association for the Study of Animal Behaviour (ASAB) in Europe and its sister organisation in the United States, the Animal Behaviour Society (ABS), in deciding whether submitted papers should be published in the journal *Animal Behaviour*. In 1981, a jointly agreed set of guidelines for the use of animals in research was produced (*Animal Behaviour*, **29,** pp 1 - 2) and over the years the editors of *Animal Behaviour* have declined, on ethical grounds, to publish a number of papers that would otherwise have been accepted for publication (Dawkins and Gosling, n.d.).

Perhaps a hypothetical example of how a cost/benefit approach could operate in deciding whether or not animals should be used for a particular research project might help specify what is involved. Consider the disease malaria. At present some two to three million people die from malaria each year and some hundred million people suffer from it but survive. In research on the prevention and cure of diseases, animals are generally used at some point in the development of vaccines and are almost invariably used in the development of new drugs. The scale of the malaria problem and the suffering caused to humans are such that a cost/benefit analysis that simply took as its goal 'minimise the total amount of suffering (both animal and human) in the world' would almost certainly allow large numbers of animals to be used in the research, even if the probability of a particular piece of research proving successful was adjudged to be quite small (Reiss, 1993a).

It is important here to note that the criterion 'minimise the total amount of suffering in the world' attempts to allow both animal and human suffering to be taken into account. Such an approach should be welcomed by biologists for whom an evolutionary understanding of the natural world reduces the distance between ourselves and animals (Rachels, 1990).

Extrapolating the cost/benefit argument from research to education, it can be argued that *some* limited decrease in the quality of life experienced by animals as a result of their use in education may be acceptable if the educational benefits are significant. For instance, keeping small mammals and fish in schools for behavioural observations and ecological investigations can be defended on these grounds provided that the possibility of animal suffering is minimised and that the educational benefits are considerable. Standard good practice with regard to housing, feeding, heating, access to water, veterinary care and appropriate care at weekends and during school holidays should be sufficient to ensure that school animals do not suffer. Boredom can be a problem for small mammals. To some extent the problem of boredom can be reduced by ensuring that the educational benefits of the animals are maximised. A balance needs to be struck between, on the one hand, small mammals being handled or investigated too often by too many people and, on the other hand, their being ignored and left unattended and unstudied for long periods of time.

DISSECTION

Although this whole publication is concerned with living organisms in biology, it seems sensible in this chapter to say a little about dissection. Dissection is the act of cutting up a plant or dead animal in order to investigate its internal structures and, by careful anatomical exploration, to reveal its organs and tissues. Over the last twenty years the dissection of whole organisms in schools has become much less common (Dixon, 1988; Reiss and Beaney, 1992; Smith, 1992; Lock, 1993). This, of course, has been partly because of changing perceptions by many pupils and some biology teachers about whether dissection is right, and partly because of campaigning by organisations such as Animal Aid and the Royal Society for the Protection of Animals. Each of these organisations has had a policy against dissection for a number of years, and both have considerable influence on school children, many of whom support their aims.

The main arguments in favour of dissection, whether of whole organisms (*e.g.* rats, fish, earthworms, cockroaches) or parts of organisms killed for other purposes (*e.g.* sheep's hearts, sheep's kidneys, bull's eyes, pig's trotters, unfertilised hens' eggs, skeletal muscle), are as follows:

- It provides a knowledge of the internal structure of tissues, organs and whole organisms better than can be provided by alternatives (*e.g.* three-dimensional models, computer software, videos).
- It enables pupils to learn through active involvement.
- It helps pupils to consider what certain careers might entail (*e.g.* medicine, veterinary medicine, animal technician).

The main arguments against animal dissection are as follows (some of these arguments apply with less force when only parts of organisms killed for other purposes are used for dissection):

- It involves the taking of life.
- It involves the rearing and killing of animals in circumstances that may cause suffering (*e.g.* physical and mental pain in battery chickens, boredom in laboratory rats).
- It lessens respect for life and so cheapens it.
- It offends some fellow pupils.
- It may put some pupils off biology.

When Lock and Millett's sample of 468 14 and 15 year olds were asked what their attitudes to dissection were, the responses were generally negative (see Table 8·4).

Table 8·4
14-15 year olds' responses when asked for their attitudes towards dissection. (taken from Lock and Millett, 1992.)

Attitude	Number of respondents	
	Females	Males
Unnecessary	63	17
Neutral	16	48
Feel squeamish	36	11
Cruel	24	20
Wrong	26	11
Use abattoir materials	17	16
Interesting	11	19
Should not be specifically killed	13	10
Use animals that died naturally	13	10
Dislike it	5	5
Of limited value	4	6
Should have choice to do	5	0
Don't use rare animals	1	1
Don't know	6	23
No response	8	17
Uncodeable	7	9

It is also interesting to note pupils' responses to the question: "If you said to a science or biology teacher 'I don't want to cut this sheep's eye open,' how do you think they would reply? (Figure 8·1). These responses are summarised in Table 8·5.

Figure 8·1. A "mouth bubble" question exploring pupil perceptions to teacher response to a refusal to dissect.

If you said this to a science or biology teacher, how do you think they would reply?

I don't want to cut this sheep's eye open

Write the teacher's reply in the bubble

Table 8·5
14-15 year olds' perceptions of teacher response to a pupil's refusal to dissect (see Figure 8·1). (Taken from Lock and Millett, 1991.)

Pupil perception of teacher response	Number of respondents	
	Females	Males
Respect view	68	54
Offer alternative work	35	17
GCSE/Exam. requirement	26	21
Educational reason	24	11
Persuade	23	21
Watch someone else	28	14
Question pupil	4	16
Sanction	8	13
Threat	4	15
Command	9	7
Don't know	7	8
Other	4	12
No response	5	4

We recommend that the following practices be followed if animal dissection is undertaken in schools:

- Each school should formulate its own policy on dissection and make it freely available to pupils and parents.
- Pupils should be told that dissection is not required by any Examining Boards or Groups in the UK.
- Pupils should be told, in advance, when dissection is to take place.
- In advance of dissection, pupils should be encouraged to discuss the ethical implications of dissection in an atmosphere that allows them to develop their thinking without being afraid that their views will be ridiculed.
- Dissection of parts of organisms obtained for other purposes should, in most cases, be used in preference to the dissection of whole organisms.
- Alternative work must always be provided. Such work should, so far as is possible, be of the same intellectual worth and interest as the dissection.
- Dissection of any sort is inappropriate for children under the age of 11.
- Dissection of whole organisms should only be undertaken by students over the age of 16. Little is gained by a teacher demonstrating a dissection to a whole class unless this is to instruct pupils or students prior to their doing their own dissection.

CONCLUSIONS AND IMPLICATIONS FOR TEACHERS

Perhaps the single most important message is that pupils should be encouraged to consider the ethical implications of their attitudes and behaviours with regard to the use of animals, and of the attitudes and behaviours of others. Such discussions in classrooms may be considered controversial. An appropriate definition of a controversial issue is provided by Dearden who writes that 'A matter is controversial if contrary views can be held on it without those views being contrary to reason' (Dearden, 1984.). Most people would accept that the issue of how humans should use animals is controversial and appropriate for discussion in school.

There are a number of different ways in which controversial issues can be introduced in science teaching (Reiss, 1993b). One approach is that of 'procedural neutrality' (Bridges, 1986). Here the teacher acts as a facilitator. Information about the controversy and different points of view are elicited from pupils and from resource material.

This approach has many advantages but obtaining suitable resource material requires a considerable input of time by the teacher. Without these appropriate resources, the approach runs the risk of failing to elicit a sufficient range of views from the pupils, in which case the lesson may become unbalanced or require the teacher to intervene. There is now a growing number of organisations that can be approached for resources to be used in classrooms on the issue of the use of animals in research and education (Animal Aid; Animals in Medicines Research Information Centre; Biomedical Research Eduction Trust; Doctors in Britain Against Animal Experiments; Research Defence Society; Dixon, 1988; Association of the British Pharmaceutical Industry, 1990; Brown, 1993). However, of these only Dixon (1988) presents views from both those in favour and those against the use of animals for human ends. It is probably best for teachers to obtain and use resources from a variety of organisations and publishers, so as to ensure a range of viewpoints.

Much of this chapter may be read as being defensive in tone. It may be argued that what we are doing is reacting to criticisms of the way animals have all too often been used, both in education and elsewhere. We feel it is important that, as considered in the other chapters of this book, the positive reasons for using animals in education are appreciated. The moral and ethical issues about the use of animals in education are not all concerned with animal suffering. There are also issues to do with enabling pupils and students to achieve a rich and full education. Biology is about the study of life. It is possible for biology teaching to enrich the education of people of all ages by using living organisms and the natural environment without causing harm.

REFERENCES

Association of the British Pharmaceutical Industry (1990) *Finding out about Medicines and Drugs*. Cambridge: Hobsons.

Bateson P. (1986) When to experiment on animals. *New Scientist*, 20 February, pp 30 - 32.

Bateson P. (1991) Assessment of pain in animals. *Animal Behaviour*, **42**, pp 827 - 839.

Bateson P. (1992) Do animals feel pain? *New Scientist*, 25 April, 30 - 33.

Bentham J. (1789) Principles of morals and legislation. In *The Collected Works of Jeremy Bentham, vol 2.1* (ed. Burns J.H. and Hart H.L.A.). London: Athlone Press. pp 11 - 12.

Bridges D. (1986) Dealing with controversy in the curriculum: a philosophical perspective. In: *Controversial Issues in the Curriculum* (ed. Wellington J.J.). Oxford: Basil Blackwell. pp 19 - 38.

Brown P. (1993) *The Search for Health: Medical Research in Action*. Cambridge: Hobsons.

Carruthers P. (1992) *The Animals Issue: Moral Theory in Practice*. Cambridge: Cambridge University Press.

Dawkins M.S. (1980) *Animal Suffering: the Science of Animal Welfare*. London: Chapman and Hall.

Dawkins M.S. and Gosling M. (n.d.) *Ethics in Research on Animal Behaviour*. London: Academic Press.

Dearden R.F. (1984) *Theory and Practice in Education*. London: Routledge and Kegan Paul. p 86.

Dixon A. (ed.) (1988) *Issues Surrounding the use of Animals in Science Lessons: a Resource Pack*. Hatfield, Herts: Association for Science Education.

Dunayer E. (1992) Animal rights issue revisited. *American Biology Teacher*, **54** pp 330-331.

Furnham A. and Pinder A. (1990) Young people's attitudes to experimentation on animals. *The Psychologist*, **10**, pp 444 - 448.

Garner R. (in press) *Animals, Politics and Morality*. Manchester: Manchester University Press.

Humphrey N. (1986) *The Inner Eye*. London: Faber and Faber.

Lock R. (1993) Use of animals in schools - pupil knowledge, experience and attitudes. In: *Ethical Issues in Biomedical Science: Animals in Research and Education* (ed. Anderson D., Reiss M. and Campbell P.). London: Institute of Biology. pp 69 - 87.

Lock R and Millett K. (1991) *The Animals and Science Education Project 1990-91: Project Report*. Birmingham: University of Birmingham.

Lock R. and Millett K. (1992) Using animals in education and research - student experience, knowledge and implications for teaching in the National Science Curriculum. *School Science Review*, **74**(266), pp 115 - 123.

Midgley M. (1983) *Animals and why they Matter: a Journey Around the Species Barrier*. Harmondsworth, Middlesex: Penguin.

Millet K. and Lock R. (1992) GCSE students' attitudes towards animal use: some implications for biology/science teachers. *Journal of Biological Education*, **26**(3), pp 204 - 208.

Rachels J. (1990) *Created from Animals: the Moral Implications of Darwinism*. Oxford: Oxford University Press.

Regan T. (1984) *The Case for Animal Rights*. Berkeley: University of California Press.

Reiss M.J. (1993a) An ethical framework for the use of animals in research and education. In: *Ethical Issues in Biomedical Science: Animals in Research and Education* (ed. Anderson D., Reiss M. and Campbell P.). London: Institute of Biology. pp 3 - 8.

Reiss M.J. (1993b) *Science Education for a Pluralist Society*. Milton Keynes: Open University Press.

Reiss M.J. and Beaney N.J. (1992) The use of living organisms in secondary school science. *Journal of Biological Education*, **26**(1), pp 63 - 66.

Singer P. (1975) *Animal Liberation: Towards an End to Man's Inhumanity to Animals*. Wellingborough, Northamptonshire: Thorsons.

Smith B. (1992) Animal use and dissection in schools - ethics, legislation and alternatives. *Lab Talk*, April, pp 30 - 33.

Smith J.A. and Boyd K.M. (Ed.) (1991) *Lives in the Balance: the Ethics of Using Animals in Biomedical Research. The Report of a Working Party of the Institute of Medical Ethics*. Oxford: Oxford University Press.

Stanisstreet M., Spofforth N and Williams T. (1993) Attitudes to children to the uses of animals. *International Journal of Science Eduction*, **15**, pp 411 - 425

Tudge C. (1988) The logical case for animal rights. *New Scientist*, 29 October, pp 6 - 7.

USEFUL ADDRESSES

Biomedical Research Education Trust,
58 Great Marlborough Street,
London   W1V 1DD.

Doctors in Britain Against Animal Experiments,
104B Weston Park,
London   N8 9PP.

Animal Aid,
7 Castle Street,
Tonbridge   YN9 1BH.

Animals in Medicines Research Information Centre,
12 Whitehall,
London   SW1A 2DY.

Research Defence Society,
58 Great Marlborough Street,
London W1V 1DD.